Algorithm Management Optimization

(Volume 1)

Artificial Intelligence in Wind Energy: Transforming Green Energy Generation

Edited by

Sugandha Singh
School of Engineering and Technology
Noida International University
Greater Noida, Uttar Pradesh
India

Inam Ul Haq
School of Engineering and Technology (SET)
CGC University, Mohali, Punjab
India

Sachin Ahuja
Department of Engineering
Chandigarh University, Mohali, Punjab
India

Akib Mohi Ud Din Khanday
Department of Information Technology
Cluster University of Srinagar
Jammu and Kashmir
India

&

Neha Sharma
Department of Computer Science and Engineering
School of Engineering and Technology Shree Guru Gobind
Singh Tricentenary University
Gurugram, India

Algorithm Management Optimization

(Volume 1)

Artificial Intelligence in Wind Energy: Transforming Green Energy Generation

Editors: Sugandha Singh, Inam Ul Haq, Sachin Ahuja, Akib Mohi Ud Din Khanday & Neha Sharma

ISBN (Online): 979-8-89881-504-2

ISBN (Print): 979-8-89881-505-9

ISBN (Paperback): 979-8-89881-506-6

Published by Bentham Science Publishers Pte. Ltd. Singapore,

First published in 2026.

need for a court order if at any point you breach any terms of this License Agreement. In no event will any delay or failure by Bentham Science Publishers in enforcing your compliance with this License Agreement constitute a waiver of any of its rights.

3. You acknowledge that you have read this License Agreement, and agree to be bound by its terms and conditions. To the extent that any other terms and conditions presented on any website of Bentham Science Publishers conflict with, or are inconsistent with, the terms and conditions set out in this License Agreement, you acknowledge that the terms and conditions set out in this License Agreement shall prevail.

Bentham Science Publishers Pte. Ltd.
No. 9 Raffles Place
Office No. 26-01
Singapore 048619
Singapore
Email: subscriptions@benthamscience.net

BENTHAM SCIENCE

CONTENTS

FOREWORD

As the world accelerates toward a sustainable future, renewable energy has become central to addressing the twin challenges of climate change and energy security. Among these, wind energy stands out as a key contributor to the clean energy mix, converting natural wind resources into electricity without harmful emissions. Yet, the inherent variability and complexity of wind patterns demand advanced solutions to ensure efficiency, reliability, and cost-effectiveness. Artificial Intelligence in Wind Energy: Transforming Green Energy Generation explores the powerful convergence of artificial intelligence (AI) and wind energy—a synergy that is redefining the future of renewable energy systems. Leveraging AI-driven decision-support tools and machine learning algorithms, wind energy operations can now be optimized across the entire value chain: from resource assessment and site selection to turbine performance optimization, predictive maintenance, and seamless grid integration.

This volume offers in-depth discussions, technical insights, and practical case studies that demonstrate how AI enhances predictive analytics, real-time monitoring, and operational efficiency. These advancements deliver tangible benefits, including higher energy yields, reduced downtime, lower maintenance costs, and the accelerated adoption of hybrid renewable systems. The book also addresses the ethical dimensions of algorithmic management, advocating for transparent, responsible, and equitable AI deployment in the energy sector. Designed for researchers, industry professionals, policymakers, and energy enthusiasts, this work provides both a comprehensive overview of current achievements and a forward-looking perspective on emerging trends and challenges. It aims to inspire innovation, foster collaboration, and contribute to the global transition toward smarter, cleaner, and more resilient energy systems.

Pramod Singh Rathore
Department of Computer and Communication Engineering
Manipal University Jaipur
Jaipur, Rajasthan
India

PREFACE

In the era of renewable energy, wind power stands as a key player in the global push for sustainable energy solutions. "Artificial Intelligence in Wind Energy: Transforming Green Energy Generation" explores how Artificial Intelligence (AI) is revolutionizing wind energy systems, improving efficiency, and driving the transition to cleaner energy sources. This book delves into the powerful synergy between AI and wind energy, offering practical insights and real-world examples of AI's impact on the wind energy sector. From resource assessment to performance optimization and grid integration, this book covers how AI-driven decision-making tools and machine learning models for resource optimization enhance every aspect of wind energy generation. Readers will discover how predictive analytics in operations management enables real-time monitoring, efficiency improvements, and cost reduction. The book also discusses ethical considerations in algorithmic management, ensuring that AI is implemented responsibly in renewable energy applications. Practical examples, case studies, and applications in real-world scenarios illustrate AI's transformative potential in areas such as wind prediction, site selection, turbine maintenance, and hybrid energy systems. Additionally, the chapters explore the economic and environmental benefits of AI, alongside emerging trends and challenges in the field.

Sugandha Singh
School of Engineering and Technology
Noida International University
Greater Noida, Uttar Pradesh
India

Inam Ul Haq
School of Engineering and Technology (SET)
CGC University, Mohali, Punjab
India

Sachin Ahuja
Department of Engineering
Chandigarh University, Mohali, Punjab
India

Akib Mohi Ud Din Khanday
Department of Information Technology
Cluster University of Srinagar
Jammu and Kashmir
India

&

Neha Sharma
Department of Computer Science and Engineering
School of Engineering and Technology
Shree Guru Gobind Singh Tricentenary University
Gurugram, India

List of Contributors

Anand Kumar Gupta	School of Computer Science in Emerging Technologies, Noida Institute of Engineering and Technology (NIET), Greater Noida, Uttar Pradesh, India
Arushi Aggarwal	Department of Computer Science and Engineering (CSE), Faculty of Engineering and Applied Technology (FEAT), Shree Guru Gobind Singh Tricentenary (SGT) University, Gurugram, Haryana, India
Anzah Bashir	School of Computer Science and Technology, Lovely Professional University, Phagwara, Punjab, India
Himani Uppal	Department of Computer Science and Engineering (CSE), Shree Guru Gobind Singh Tricentenary University (SGT) University, Gurugram, Haryana, India
Harsh Kumar	Department of Computer Science & Engineering, Akal University, Talwandi Sabo, Punjab, India
Hashim Mohi u din Dar	Department of Computer Science & Engineering, Rayat Bahra University, Mohali, Punjab, India
Inam Ul Haq	School of Engineering and Technology (SET), CGC University, Mohali, Punjab, India
Inderdeep Kaur	Department of Computer Science & Engineering, University Institute of Engineering, Chandigarh University, Mohali, Punjab, India
Inzimam Ul Hassan	Department of Computer Science & Engineering, Vivekananda Global University, Jaipur, India
Khalid Hafiz Mir	School of Computer Science and Technology, Lovely Professional University, Phagwara, Punjab, India
Mamta	Computer Science and Engineering (CSE), Chandigarh University, Mohali, Punjab, India
Manik Rakhra	School of Computer Science and Engineering, Lovely Professional University, Phagwara, Punjab, India
Malik Muzamil Ishaq	Department of Computer Science & Engineering, Akal University, Talwandi Sabo, Punjab, India
Shivani	Department of Computer Science & Engineering, Indira Gandhi University, Rewari, Haryana, India
Satinder Bal Gupta	Department of Computer Science & Engineering, Indira Gandhi University, Rewari, Haryana, India
Sakshi	Department of Computer Science and Engineering (CSE), Faculty of Engineering and Applied Technology (FEAT), Shree Guru Gobind Singh Tricentenary (SGT) University, Gurugram, Haryana, India
Sheetal	Department of Computer Science and Engineering (CSE), Shree Guru Gobind Singh Tricentenary University (SGT) University, Gurugram, Haryana, India
Syed Zoofa Rufai	Department of Computer Science and Engineering (CSE), Shree Guru Gobind Singh Tricentenary University (SGT) University, Gurugram, Haryana, India
Tiyas Sarkar	School of Computer Science and Engineering, Lovely Professional University, Phagwara, Punjab, India

Vandita Nandal	Department of Computer Science and Engineering (CSE), Shree Guru Gobind Singh Tricentenary University (SGT) University, Gurugram, Haryana, India
Vikas Verma	School of Computer Science and Engineering, Lovely Professional University, Phagwara, Punjab, India

CHAPTER 1

Introduction to AI in Wind Energy

Mamta[1,*]

[1] *Computer Science and Engineering (CSE), Chandigarh University, Mohali, Punjab, India*

Abstract: Wind energy is transformed using Artificial Intelligence technology, resulting in improved system performance alongside efficient operations. The fundamental functions by which AI systems enhance wind energy technology form the focus of this chapter. This chapter illustrates the development of AI technology into important resource evaluation systems and weather forecast applications for wind power operations. Monitored data patterns through AI technologies enable better operational efficiency of the wind power grid. Artificial Intelligence systems improve wind energy operations by increasing system performance, which simultaneously minimizes operational expenses. Through AI integration, connectivity problems in wind-solar hybrid renewable systems are solved, along with enhanced operational reliability. This chapter analyzes both the practical sustainability challenges of AI implementation as well as its essential advantages for human beings and Earth. This chapter examines the latest technological developments in Artificial Intelligence applications. This chapter provides details on how AI technology drives wind power system advancements toward sustainable renewable energy development.

Keywords: Artificial Intelligence, Hybrid wind-solar integration, Mesoscopic methods, Predictive analytics, Renewable energy systems, Sustainable energy solutions, Wind energy optimization.

INTRODUCTION

The modern implementation of AI within wind energy continues to advance renewable technology by transforming the procedures for wind turbine management as well as wind resource optimization. Sophisticated algorithms with machine learning capabilities, along with data analytical methods, enhance wind energy system performance by increasing efficiency and reducing operating expenses [1].

Scientists now emphasize wind energy adoption because it represents a sustainable electricity generation method required to fulfill our increasing demand

* **Corresponding author Mamta:** Computer Science and Engineering (CSE), Chandigarh University, Mohali, Punjab, India; E-mail: mamta.airesearch@gmail.com

Sugandha Singh, Inam Ul Haq, Sachin Ahuja, Akib Mohi Ud Din Khanday & Neha Sharma (Eds.)

for eco-friendly power sources. Modern wind-power system enhancements are accelerating as artificial intelligence provides solutions to common issues affecting wind-based energy generation and management.

DEFINITION AND IMPORTANCE OF AI IN RENEWABLE ENERGY

For renewable energy applications, Artificial Intelligence represents computer systems that show high performance at tasks requiring human mental skills. Through their ability to learn autonomously, such systems adapt to changes in input data while completing complex humanlike tasks without continuous supervision [2]. In wind energy specifically, AI encompasses several key technologies:

i. **Machine Learning Systems:** These systems evaluate large datasets that contain wind turbine data alongside weather patterns and grid requirements to produce predictions and optimized operations. These systems see connections between data that human operators cannot observe independently.
ii. **Neural Networks:** Complex computational systems that follow brain-based designs analyze advanced datasets and generate stand-by operations choices linked to turbine maintenance requirements, together with power output predictions.
iii. **Predictive Analytics:** Technology enables predictions of future weather patterns and energy requirements, along with machine failure detection, to help actively manage wind power facilities.

The importance of AI in renewable energy, particularly wind power, can be understood through several critical aspects:

Enhanced Efficiency

- The turbine system achieves maximum efficiency through real-time adjustments of blade positions, combined with rotational velocity optimization based on current wind conditions.
- Predicting weather pattern changes and implementing them enables systems to enhance their electricity production capabilities.
- Energy losses are reduced because of the smart integration and management capabilities of smart grids.

Cost Reduction

- Minimizes operational costs through predictive maintenance.
- Predictive maintenance reduces operational expenditures because of early equipment observations.

- Forecasting utilization minimizes operational shutdowns by detecting predictable equipment malfunctions before failure points.
- The system helps organizations deploy their workforce optimally while allocating their resources most efficiently.

Grid Integration

- The power distribution grid achieves increased stability as a result of this integration.
- Integrating this system helps enhance the management of irregular power output as it occurs.
- The system promotes seamless integration between renewable energy sources along with other renewable energy sources.

EVOLUTION OF AI APPLICATIONS IN THE ENERGY SECTOR

Artificial Intelligence integration in the energy sector is an exciting technological development [3]. Enhancing basic operational efficiency became the primary objective of the implementation. The field advances with additional sophisticated monitoring equipment coupled with modern sensing systems. Experts have developed the first AI applications for the energy sector in the past decades using artificial human decision-support systems. The initial energy applications served to monitor and control energy system operations. The first applications of Artificial Intelligence served the purpose of energy system control and monitoring. The industry experienced its first true revolution through Machine Learning and Deep Learning when they emerged. These technologies facilitated the analysis of large datasets, resulting in significant advancements in predictive maintenance, energy forecasting, and grid management. Fig. (**1**) illustrates the history of AI and its integration through three stages: Early, Intermediate, and Modern Era. Basic monitoring systems and automated controls were introduced in the Early Phase (1990s–2000s), whereas predictive maintenance and data storage were introduced in the Intermediate Phase (2000s–2015), along with machine learning algorithms. Modern Era (2015–Present): This includes advanced AI methodologies like deep learning, real-time analytics, smart grids, IoT, and cybersecurity. This progression can be traced through several distinct phases:

i. Early Development (1990s - 2000s)
   - Initial implementation of basic monitoring systems.
   - Simple automated control mechanisms.
   - Limited data collection and analysis capabilities.
   - Focus on basic operational efficiency improvements.

ii. Intermediate Phase (2000s - 2015)
   - Introduction of more sophisticated sensors and monitoring equipment.
   - Development of basic predictive maintenance systems.
   - Implementation of preliminary machine learning algorithms.
   - Enhanced data collection and storage capabilities.
   - Integration of weather forecasting systems.

iii. Modern Era (2015 - Present)
   - Advanced machine learning and deep learning applications.
   - Real-time data analytics and decision-making capabilities.
   - Sophisticated predictive maintenance systems.
   - Integration with Internet of Things (IoT) devices.
   - Development of smart grid technologies.
   - Enhanced cybersecurity measures.

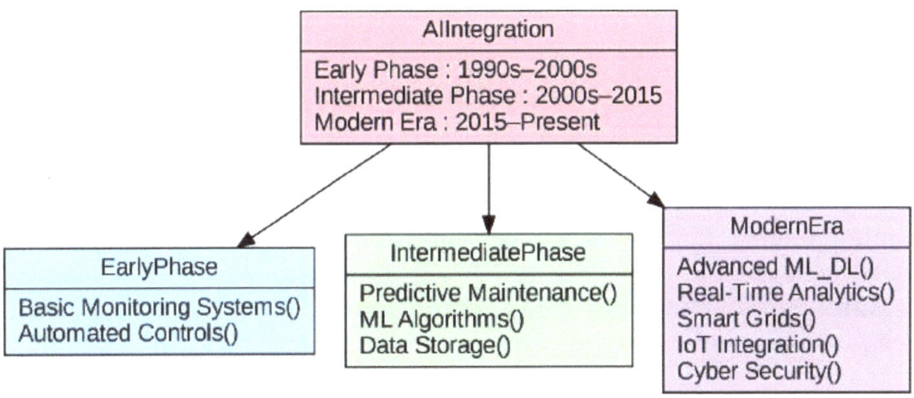

Fig. (1). Evolution of AI in the energy sector.

CURRENT APPLICATIONS OF AI IN THE ENERGY SECTOR

AI has become deeply involved in multiple operations spanning energy production and distribution, and use processes. A brief overview of the transformation of artificial intelligence in the energy sector is presented in Fig. (**2**), with a clear graphical structure. The central concept of "AI in the Energy Sector" distributes its five fundamental applications across vertical sections. The text mentions Predictive Maintenance—using AI to predict moments of equipment failure and reduce overhead—Demand Forecasting, optimizing power supply and cutting costs with accurate predictions of electricity consumption—Smart Grids, integrating renewable energy sources with AI to ensure constant reliability—Energy Efficiency, monitoring consumption and operational processes for better resource management—and Digital twins, generating virtual models to simulate and optimize the performance of assets. Here are a few important applications:

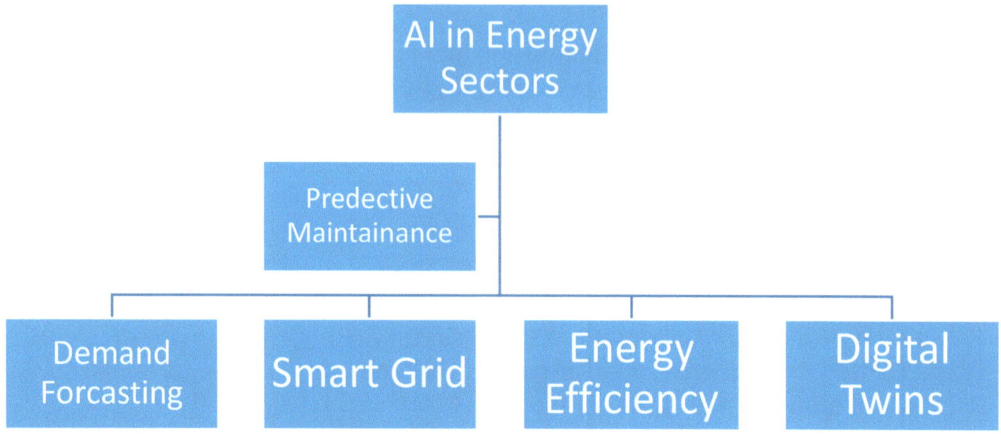

Fig. (2). Key AI applications transforming the energy sector.

i. **Predictive Maintenance and Asset Management:** AI-based predictive maintenance is critical to reducing downtime and optimizing operational costs. AI can analyze sensor data, compare it with familiar temperature fluctuations, and reference historical performance records to forecast when equipment is likely to fail. This allows better scheduling of maintenance during optimum times, reducing unplanned outages and increasing the lifespan of critical infrastructure.
ii. **Demand Forecasting and Load Management:** Smart grids use AI, cloud tech, and digital solutions to optimize energy distribution and load management. AI algorithms are used to optimize grid operations by predicting how much energy will be generated from natural sources such as sunlight and wind, and by better integrating them into the power grid. This integration enhances grid stability and efficiency, ensuring a reliable electricity supply.
iii. **Smart Grids and Renewable Energy Integration:** Smart grids leverage AI, cloud technologies, and digital solutions to enhance energy distribution and load management. AI algorithms optimize grid operations by predicting energy output from renewable sources like solar and wind, enabling better integration into the power grid. This integration improves grid stability and efficiency, ensuring a reliable electricity supply.
iv. **Energy Efficiency Programs:** AI plays a vital role in energy efficiency by gathering data from building systems, sensors, and external sources to identify unnecessary expenses and predict peak consumption periods. Machine Learning systems modulate HVAC, lighting, and other processes to match real-time conditions, optimizing energy usage in line with real-world demand rather than pre-programmed schedules.
v. **Digital Twins and Monitoring:** Digital twins are the virtual representations of physical assets that enable remote monitoring and control. With

more federated data, engineers take an AI-driven digital twin to simulate different scenarios, which will allow them to explore potential issues or optimize the performance of assets without the risks of costly experimentation.

Real-world Applications of AI in Energy

Many companies have successfully integrated AI into their operations, proving its ability to revolutionize the energy sector:

i. **Engie Energy Access:** Engie Energy Access applied AI to find potential solar panel customers across sub-Saharan Africa. Using machine learning and geospatial data, they developed targeted marketing plans and dramatically increased sales.
ii. **Duke Energy:** Duke Energy used AI for monitoring and mitigation of methane emissions from natural gas pipelines. AI processed data from sensors and satellites to locate leaks, allowing for quick repairs and fewer greenhouse gas emissions.
iii. **Marathon Oil:** At Marathon Oil, AI was applied to time-series data from the company's wells to build intelligent alerts on changes in conditions. This dramatically reduced response times and increased operational efficiency.

Key Evolutionary Milestones

i. **Data Collection and Analysis:** The progression from basic initial monitoring tools toward modern, real-time, sophisticated analytics solutions has marked the evolutionary path for this technology [4]. Modern operational systems process wind farm data obtained from various inputs to generate extensive information about their wind farm performance.
ii. **Predictive Capabilities:** Computers based on early systems processed simple statistical algorithms, but current artificial intelligence technologies apply cutting-edge machine learning frameworks to deliver precise predictions about:
   - Wind patterns and power generation potential.
   - Equipment maintenance needs.
   - Grid demand fluctuations.
   - Optimal operational parameters.
iii. **Automation and Control:** Wind farm management shifted toward full automation through AI-driven operations after transitioning from manual control systems. Modern systems can:
   - The system can automatically modify turbine components to reach peak operational effectiveness

- The system merges multiple turbines into optimal functioning complexes.
- Operators use real-time data processing to meet grid requirements.
- The system controls both power storage management and distribution processes.

iv. **Integration and Connectivity:** The evolution of AI applications has led to increasingly interconnected systems:

- Advancements in artificial intelligence applications have generated many interlinked operational systems.
- Turbine units now connect through enhanced communication networks.
- Modern systems integrate with weather forecasting systems for data analysis.

v. **Safety and Reliability:** AI development has brought considerable advances to safety and reliability because of:

- Enhanced monitoring capabilities.
- Better risk assessment.
- Improved emergency response systems.
- More accurate failure prediction.

The evolution continues today with emerging technologies and applications:

i. Advanced Analytics

- Implementation of quantum computing capabilities.
- Development of more sophisticated AI algorithms.
- Enhanced pattern recognition capabilities.
- Improved decision-making processes.

ii. Smart Grid Integration

- Better coordination with other power sources.
- Enhanced demand response capabilities.
- Improved grid stability management.
- More efficient energy distribution.

ROLE OF AI IN ENHANCING WIND ENERGY SYSTEMS

AI technologies implemented in wind energy systems have transformed operational processes, achieving new heights in system reliability and operational efficiency in this sector [5]. Advances in these wind power systems function as the fundamental structure of present-day wind farms through their capacity to automate smart decisions and maximize asset utilization.

Resource Assessment and Wind Prediction

i. **Advanced Weather Modeling:** Modern AI systems transform weather modeling for wind energy applications by combining multiple data sources with sophisticated algorithms [5]. The key components include:
   - The integration of neural networks with established numerical weather prediction models enhances prediction accuracy while extending operational horizons.
   - The system accomplishes resource assessment through satellite imagery, ground sensors, and historical data patterns analysis.
   - Real-time processing monitors forecast evaluations through weather condition modifications.

ii. **Wind Pattern Analysis:** Neural networks enabled a remarkable development of wind pattern analysis for extending wind resource assessment from short to long-term perspectives. This technology provides crucial insights through:
   - The combination of computational fluid dynamics and machine learning achieves advanced turbulence modeling.
   - Location studies of wind patterns, both geographic and seasonal, enable effective farm placement optimization.
   - The system employs long-term pattern identification techniques to find productive operating points and forecast future energy output levels.

iii. **Power Generation Forecasting:** AI systems function at various time scales to provide power generation predictions that support three different operational requirements. Critical aspects include:
   - Operational choices together with market entry require short-term forecasting spanning between zero and forty-eight hours.
   - The automated system performs long-range forecasting to support investment planning while helping organizations decide about capacity expansions.

iv. **Site Optimization:** The extensive approach to wind farm development and operation that stems from artificial intelligence applications serves as site optimization AI [6]. Key features include:
   - The software features sophisticated terrain analysis tools for precise turbine site placement and position strategies that lessen turbine effects.
   - The combination of environmental impact assessments should operate within performance optimization models.
   - Operational data, together with changing environmental conditions, drive dynamic changes to site layouts.

Real-time Monitoring and Data Analytics

i. **Sensor Integration and Data Collection:** Modern wind farms integrate advanced sensor networks that collect data from both specific turbine measurements and environmental conditions [7]. Essential elements include:
   - Comprehensive turbine monitoring systems track mechanical, electrical, and thermal parameters.
   - Environmental monitoring stations collect atmospheric and weather-related data.
   - Advanced data acquisition systems with real-time processing capabilities.

ii. **Performance Analytics and Health Monitoring:** The integration of AI in performance analytics has revolutionized wind farm operations through:
   - The use of predictive maintenance algorithms that enables the system to predict equipment failures before their actual occurrence.
   - Automated turbine parameter control that comes from real-time optimization systems with built-in performance analysis.

iii. **System Integration and Grid Management:** Power quality management systems alongside dynamic load balancing operate through the platform and the power grid through:
   - Dynamic load balancing and power quality management systems.
   - Real-time grid stability analysis and automatic response mechanisms.
   - Intelligent storage system integration for optimal power delivery.

iv. **Data Processing and Decision Support:** Current wind operations produce copious amounts of data, which demands processing architectures designed for decision support tasks [8]. Key features include:
   - Real-time data validation and anomaly detection systems.
   - Advanced analytics platforms for historical trend analysis and performance optimization.
   - Maintenance operation planning receives automated support through computer systems to make decisions.

v. **Visualization and Operational Enhancement:** The system includes automated reporting, enabling customizable alert systems with notification capabilities. Advanced visualization tools transform complex data into actionable insights through:
   - Interactive dashboards displaying real-time performance metrics and system status.
   - Automated reporting systems with customizable alerts and notifications
   - Predictive analytics visualizations for maintenance planning and resource allocation.

AI FOR PREDICTIVE MAINTENANCE OF WIND TURBINES

Wind energy is one of the critical elements of the global shift towards renewable energy, providing a sustainable and eco-friendly alternative to fossil fuels. Nevertheless, wind turbines face a multitude of challenges in terms of efficiency and reliability due to harsh environmental conditions as well as the complexity of their mechanical structures and the associated high maintenance costs. Traditional maintenance methods, such as reactive or scheduled maintenance, frequently result in unanticipated failures, expensive downtime, and ineffective resource usage. AI-driven predictive maintenance provides a revolutionary solution by combining machine learning (ML), deep learning, and data acquisition to anticipate failures before they occur. With AI integration into wind turbine maintenance, operators can improve the performance of wind energy assets, reduce downtime, and extend their lifespan.

The Need for Predictive Maintenance in Wind Turbines

Wind turbines are constantly exposed to a multitude of external and internal stresses, such as severe weather conditions (above and below), changing wind speeds, temperature changes, humidity, and mechanical stresses on moving parts. These factors eventually lead to wear and tear on important components like blades, gearboxes, and generators. Unplanned failures are expensive to repair and can also halt power production and disrupt energy distribution. The conventional maintenance strategies comprise:

- **Reactive Maintenance:** In this approach, assets are repaired or replaced only after a failure occurs. The cheapest solution is the most expensive solution — it may appear like it's less expensive at first, but reactive maintenance leads to a scenario of downtime, high repair costs, and unsafc conditions.
- **Preventive Maintenance:** This is usually carried out at regular intervals and helps minimize the risk of failures. What that means is that, under this method, an asset could be inspected when it does not need to be, or a part of it might even be replaced at an early stage, adding costs to ops.

These limitations are mitigated through AI-driven Predictive maintenance, which leverages real-time sensor information, past performance data, and sophisticated algorithms to forecast possible component breakdowns and fine-tune maintenance schedules accordingly. Through AI-powered predictive maintenance, wind farm operators can take proactive steps to prevent potential breakdowns, ultimately minimizing downtime and operational expenses while maximizing efficiency.

AI TECHNIQUES USED IN PREDICTIVE MAINTENANCE

Machine Learning Algorithms

This analysis is carried out using machine learning techniques that operate on large datasets from turbine sensors to find patterns and predict failures before they occur. Some key ML methods include:

- **Supervised Learning:** Training AI models using labelled historical data, where the normal and faulty states of turbine components are clearly defined. Classification and failure prediction can be performed using algorithms such as decision trees, support vector machines (SVMs), random forests, *etc.*
- **Unsupervised Learning:** This method is applied for anomaly detection in the absence of any labeled data. Algorithms such as k-means and principal component analysis (PCA) are used to cluster and identify abnormal behavior in turbine performance.
- **Reinforcement Learning:** Real-time turbine data enables AI models to learn in every possible way and update maintenance schedules on the fly to improve efficiency and reduce operational risk.

Deep Learning Models

Predictive maintenance is enhanced using deep learning because it enables the identification of complex patterns in sensor data, as well as enhances failure detection accuracy. Deep learning techniques include:

- **Convolutional Neural Networks (CNNs):** CNNs can take image inputs, making them ideal for fault detection based on images such as surface cracks and thermal or infrared images of turbine blades.
- **Recurrent Neural Networks (RNNs) and Long Short-Term Memory (LSTM):** Useful for time-series data analysis, capturing trends, and gradual degradation patterns in turbine components.
- **Autoencoders:** Used for anomaly detection, learning the normal operating behavior of the turbines, and flagging deviations that indicate possible failures.

Digital Twin Technology

Digital twins, are virtual mirrors of their real-world counterparts that constantly feed sensor data from their physical wind turbines into this virtual environment. AI-driven digital twins have many benefits:

- Simulation of relevant conditions to determine component performance and failure predictions.
- Predicting failures and testing for scenarios without physical intervention, and thus minimizing maintenance risks.
- AI-powered cycle optimization across different processes in manufacturing, increasing the efficiency of equipment or the plant as a whole.

Internet of Things (IoT) and Edge Computing

Turbines are equipped with Internet-of-Things (IoT) sensors that collect huge amounts of operational, real-time data like temperature, vibration, pressure, wind speed, and generated power. Machine learning tools, run in edge-computing environments by AI, process this data, which streamlines latency, minimizes lag, and quickens response time when predictive maintenance alerts are triggered. AI systems can run live analytics on IoT data to monitor equipment and detect warning signs of equipment failures, equipping operators to take preventive measures.

KEY COMPONENTS OF AI-BASED PREDICTIVE MAINTENANCE SYSTEMS

Data Collection and Processing

Various sensors in wind turbines continuously capture operational data. Some of the important parameters that are monitored are:

- **Vibration Sensors:** Monitor rotating parts like gearboxes and bearings for abnormalities.
- **Temperature Sensors:** Detect overheating problems in key components.
- **Acoustic Sensors:** Detect abnormal noise patterns, which could indicate mechanical malfunctions.
- **Supervisory Control and Data Acquisition (SCADA) Systems:** Group turbine data for centralized monitoring and control.

Feature Extraction and Anomaly Detection

In the case of detecting faulty components in gas turbines, AI models learn to recognize the nuances of what constitutes "normal" versus "abnormal" behavior by sifting key feature information from raw sensor input. Some of the types of anomaly detection techniques are:

- **Statistical Methods:** Find anomalies from normal operating ranges.
- **Cluster Analysis:** Like a satellite, it keeps an eye on redundant data points to detect outliers that can signal an impending failure.
- **Time-Series Analysis:** Monitors historical trends to detect directional degradation in turbine performance.

Failure Prediction and Decision-Making

AI-powered predictive maintenance systems predict failures and offer actionable recommendations for potential failures. Decision models use:

- **Classification Algorithms:** Classify issues as a minor, medium, or critical fault.
- **Predictive Analytics:** Estimate the remaining useful life of turbine components.
- **Prescriptive Maintenance:** Provides recommendations for best maintenance actions based on AI-driven predictions, enhancing decision-making processes.

BENEFITS OF AI IN WIND TURBINE PREDICTIVE MAINTENANCE

i. **Reduced Downtime and Increased Availability:** AI recognizes early symptoms of component failure and takes preventive measures before failures occur, significantly reducing unforeseen turbine failures and enhancing energy production.
ii. **Cost Savings on Maintenance and Repairs:** AI can optimize maintenance schedules, reducing unnecessary inspections and replacements, as well as preventing major failures that incur costly repairs.
iii. **Extended Equipment Lifespan:** Continuous monitoring and proactive maintenance help mitigate long-term wear and tear, thereby extending turbine life and improving return on investment (ROI).
iv. **Improved Energy Efficiency:** Utilizing AI for optimization ensures the turbines operate at peak performance, maximizing energy savings and reducing operational inefficiencies.
v. **Enhanced Safety for Maintenance Personnel:** AI reduces the need for manual checks and repairs, keeping workers safer from hazardous working conditions.

CHALLENGES AND LIMITATIONS

In this section, we will discuss the challenges and limitations of our approach. While beneficial, AI-based predictive maintenance has several challenges:

i. **Data Quality and Availability:** Variations in sensor data, missing entries, and noise can affect the model's ability to generalize well to unseen data and may necessitate effective data preprocessing and validation.

ii. **Integration Complexity:** The introduction of AI solutions requires compatibility with existing SCADA and IoT systems, which can entail significant initial costs and technical expertise.
iii. **Model Interpretability and Trust:** AI models can be considered "black boxes," which raises the challenge of interpreting predictions, requiring transparency and explainability techniques.
iv. **Cybersecurity Concerns:** Generative AI solutions are anchored in cloud-based data control and, therefore, are open to cyber threats and require vigorous cybersecurity efforts.

FUTURE TRENDS AND INNOVATIONS

i. **Federated Learning for Decentralized Data Processing:** Allows AI models to be trained on data generated by different turbines while keeping the archives local and secure.
ii. **Autonomous Drones for AI-Powered Inspections:** Drones with computer vision powered by AI can now remotely identify damage to blades and structural issues.
iii. **Hybrid AI Models for Enhanced Accuracy:** Integrating ML with physics-based models enhances fault detection and maintenance scheduling.
iv. **AI-Driven Robotics for Automated Repairs:** AI-operated robots may be able to handle small maintenance jobs while minimizing human reliance and operational costs.

AI-DRIVEN OPTIMIZATION OF WIND ENERGY

The implementation of artificial intelligence optimization strategies in wind energy systems yields substantial improvements in performance and profit potential [9]. Such systems that combine advanced algorithms with real-time data processing deliver better performancc than any previous system has achieved.

Improving Energy Efficiency and Output

Wind farms gain revolutionary advancements in energy efficiency and output optimization through complete management systems and progressive control approaches made possible by AI technology.

i. **Advanced Turbine Control Systems:** The operation of contemporary wind farms depends on AI-based control systems running advanced functions to optimize turbine operation throughout time [10]. The systems accomplish this *via*:
   - A system for adjusting blade pitch elements along with yaw control systems runs operations by processing real-time wind data and analyzing turbine

loads.
  - Turbine farms achieve maximum output by using smart wake-steering systems that synchronize multiple units to gradually reduce wake.
  - The technology considers advanced load management systems that optimize mechanical stress reduction alongside power delivery.

ii. **Performance Enhancement Strategies:** The AI system runs continuous assessments of operational aspects to achieve peak energy output and extend the system's lifespan.
  - The existing environmental conditions guide advanced machine learning technologies in adjusting turbine control codes.
  - The system uses neural networks to adjust operational parameters in real time, resulting in optimal power curve performance.
  - Adaptive system algorithms learn to optimize their efficiency by analyzing historical operational data [11].

iii. **Grid Integration Optimization:** AI-based grid integration optimization systems show significant enhancement in the process of connecting wind energy to power grids.
  - Modern controller technologies implement intelligent system adjustments to manage turbine power generation while maintaining grid stability requirements.
  - The system uses predictive algorithms to optimize the timing of power supply deliveries based on demand.
  - Earth Intelligence as a Framework for Optimized Grid Functionality abstracts advanced frequency and voltage control systems that stabilize grids and maximize the output of power.

COST REDUCTION STRATEGIES USING AI

Communication between the wind energy sector and AI allows the industry to develop cost-effective infrastructure that controls operational activities and maintenance operations [12].

i. **Predictive Maintenance Optimization:** The introduction of AI-driven maintenance methods completely changed how organizations handle equipment reliability during maintenance operations.
  - The system uses advanced wear-prediction models to determine optimal component lifecycle times.
  - Maintenance schedule decisions are based on risk assessments that enable interventions to target areas with the highest failure impact.
  - Advanced computers now perform diagnostic testing with greater accuracy at lower costs for inspections.

ii. **Operational Cost Management:** Operational Cost Management receives extensive cost optimization capabilities from AI systems that offer:
   - Advanced intelligent systems determine how to distribute resources, making sure every workforce requirement meets equipment efficiency needs.
   - Profit-maximizing systems work in conjunction with operational expense-reduction techniques to enable energy market participation.
   - Sophisticated inventory management tools manage spare part availability alongside affordable storage needs.

iii. **Lifecycle Cost Optimization:** Advanced life cycle cost management becomes possible because of AI:
   - Leadership enhancement software serves to balance system performance against operational lifespan duration.
   - The planning systems use evaluations to determine optimal times for upgrades and system replacement decisions.
   - Reliability-based equipment systems benefit from operational strategy optimization that uses expected equipment lifespan data.

iv. **Implementation Considerations:** A successful deployment of AI-driven optimization strategies needs several essential implementation factors that drive their implementation success.

TECHNICAL INFRASTRUCTURE REQUIREMENTS

Organizations must build AI systems on a stable technical infrastructure to achieve success. Organizations need data processing technology coupled with dependable sensors that operate over fast networks to manage real-time high-volume data flows [13]. Organizations need to guarantee their system infrastructure will handle advanced AI system requirements with reliable and secure elements.

i. **Data Management Strategy:** Effective data management is crucial for AI-driven optimization. This involves:
   - The strategy includes detailed rules for acquiring complete high-quality data through systematic processes.
   - Modern data storage systems provide simultaneous fast data retrieval alongside complete data protection and integrity.
   - Complex data processing methods convert unprocessed information to useful operational knowledge.

ii. **Return on Investment Considerations:** The implementation of AI optimization systems requires careful evaluation of potential returns:
   - The assessment for initial deployment must examine all required hardware elements, software components, and employee training programs.

- High operational costs are evaluated through two dimensions: system maintenance and program updates.

AI-driven optimization strategies demonstrate their value through improved operational outputs, which translate into reduced operational costs. Organizations implementing these systems typically observe:

- 10-20% increase in overall energy output through improved operational efficiency.
- Organizations achieve a 15-30% decrease in maintenance expenses through predictive maintenance approaches.
- These AI-driven optimization strategies lead an organization to increase its workforce utilization rate by 20-40% by using better resource management approaches [14].

The Levelized Cost of Energy reduces significantly due to these advancements, which establishes wind power as more competitive than traditional energy sources. The evolution of Artificial Intelligence technology shows a bright future for improved wind energy operation with cost reductions that will sustain its essential role in powering worldwide energy systems.

INTEGRATION OF AI IN HYBRID RENEWABLE SYSTEMS

Hybrid renewable energy systems powered by Artificial Intelligence represent a major development for efficient, sustainable energy solutions. Wind and solar energy sources working together through hybrid systems solve renewable resource reliability problems to maintain an uninterrupted power supply [15]. AI technology improves renewable systems through its analytical capabilities and predictive functions, together with its operation automation features.

Wind-solar hybrid systems

The complementary operational characteristics of wind and solar power supply allow wind-solar hybrid systems to generate maximum power. AI plays a pivotal role in optimizing these systems by addressing the following aspects:

- **Resource Optimization:** AI predicts power generation performance with accuracy by processing current weather information, which includes information regarding both wind speed and solar irradiance data [16]. The system optimizes resource management by properly distributing assets throughout different environmental conditions.
- **Dynamic Load Management:** Through dynamic intelligent programmatic control systems, adjust the energy distribution by choosing between wind and

solar energy power production according to their availability and efficiency levels. The flexible system design cuts down energy waste while delivering enhanced performance results.

- **Cost Reduction:** Through automated procedures, together with Predictive analytics, companies can lower operational expenses by extending service availability while enhancing energy system efficiency [17]. A computing-aided approach provides businesses with greater control over maintenance timelines and scheduling in hybrid energy systems.
- **Energy Storage Management:** The use of AI lets operators manage energy storage systems to avoid wasting high-output energy during peak times and provide optimum utilization during low-energy periods.
- **Grid Stability:** The control of energy input, together with storage capabilities through AI, prevents system overload and maintains consistent power delivery for building a resilient network infrastructure.
- **Sustainability:** Wind-solar systems powered by AI enable greenhouse gas reductions and fuel independence, benefiting global sustainability outcomes.

ROLE OF AI IN GRID INTEGRATION

The sporadic behaviour of wind and solar power resources creates substantial hurdles when incorporating hybrid renewable systems into the electrical grid [18]. The challenges of the grid integration of hybrid renewables and the respective AI-driven solutions are illustrated in Fig. (**3**). These challenges are grid inefficiencies, cybersecurity, scalability, faults, and energy fluctuations. AI-powered solutions can help overcome these challenges with demand prediction, improved security, fault detection, energy flow optimization, and smart grid advancements to optimize energy distribution and integration. AI addresses these challenges through:

- **Real-Time Grid Monitoring:** AI systems track grid performance nonstop, which helps them recognize operational weaknesses so they can execute instant adjustments to enhance energy delivery.
- **Demand Prediction:** The predictive capabilities of advanced AI models achieve energy demand patterns with remarkable accuracy, which supports better resource distribution as well as active prevention of energy shortages [19]. Regions with irregular energy consumption demand continue to heavily depend on predictive capabilities that forecast their patterns.
- **Energy Flow Optimization:** The system uses intelligent programming in an automated process to distribute power between generation sources and electric grid networks, which provides continuous, stable power delivery. A system must control energy distribution between high- and low-demand periods to enable effective demand response operations.

- **Fault Detection and Recovery:** Artificial intelligence enables the system to automatically detect and recover equipment faults in both power grids and hybrid systems to prevent service interruptions and maintain stable grid operation. System stability increases as operational continuity improves, since computing elements are located in central locations.
- **Renewable Energy Storage Integration:** AI monitoring technology enables efficient operation of storage systems, including batteries and pumped storage facilities, effectively managing excess energy during periods of high demand [20]. The system allocates energy supply during peak use periods without creating waste and remains accessible when essential.
- **Advanced Control Mechanisms:** AI facilitates the evolution of smart grids by networks of connected devices that can operate autonomously, share operational information without any human oversight, and build self-correcting mechanisms to ensure grid stability. By controlling energy flow, behavioral control mechanisms help balance load distribution and enhance grid operational stability.
- **Enhanced Security:** The AI-based cyber-attack detection systems help protect hybrid RES and/or grid infrastructures from attacks and enable secure, reliable energy management.
- **Scalability:** The implementation of AI enables hybrid renewable energy systems to scale their capabilities by delivering optimized procedures for meeting increased power demands.

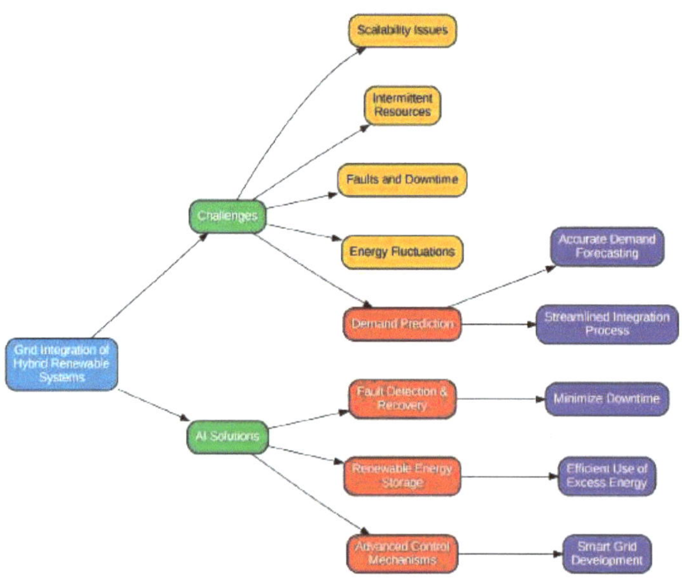

Fig. (3). Empowering renewable energy with AI: tackling challenges and unlocking solutions for a sustainable grid.

ETHICAL AND ENVIRONMENTAL CONSIDERATIONS

The implementation of Artificial Intelligence controls within renewable energy systems produces various advantages but creates moral and environmental complications. For sustainable AI technology usage in energy systems, both ethical considerations and fair operation must be addressed.

ADDRESSING CHALLENGES IN AI IMPLEMENTATION

The adoption of AI systems within renewable energy requires organizations to address various technical problems that demand thorough examination:

- **Data Privacy and Security:** Computer systems utilizing AI processing require elaborate datasets consisting of private information, which must be given special consideration for security purposes. Security protocols and data privacy practices create dual protection barriers against user information violations while preventing the occurrence of schema threats. Voice system security benefits simultaneously from testing practices combined with encryption standards for better protection.
- **Algorithmic Bias:** The training algorithms that power AI models accept data biases from their training sources to produce unfair decisions affecting specific demographic groups [21]. The algorithms that distribute energy show preferential treatment for specific geographical locations. The design of transparent, unbiased algorithms represents the primary focus for creating equitable outcomes. Essential to the market is the ongoing evaluation of Artificial Intelligence systems for both bias and inclusivity detection.
- **Energy Consumption of AI Systems:** Using Artificial Intelligence models for training alongside operational activities requires massive computational power, resulting in elevated energy consumption. The energy consumption required for renewable energy development suffers as a result of these counterproductive methods [22]. The integration of renewable energy into operations, optimized with efficient algorithms and efficient hardware, enables us to minimize this issue.
- **Access and Equity:** Every community needs equal access to the benefits Artificial Intelligence delivers to renewable power systems. To prevent the gap between developed and underdeveloped regions from widening, we need to address economic and technological inequality. The gap between renewable energy capabilities and affordability can be closed through combined efforts and strategic government policy implementation.
- **Ethical Use of AI:** Guideline and regulatory frameworks for AI in energy systems protect these technologies while respecting social values and aligning with environmental targets. AI developers will develop responsible practices

through both ethical development training protocols and clear accountability support frameworks.

ENSURING SUSTAINABILITY AND ENVIRONMENTAL PROTECTION

AI technologies have the potential to enhance sustainability efforts in renewable energy, but must be implemented with environmental protection in mind:

- **Minimizing Environmental Impact:** The design and implementation of AI systems need to focus on minimizing ecological impact by reducing resource extraction and energy consumption required by AI network infrastructure. Implementing green manufacturing strategies for building AI equipment systems will create extra value.
- **Optimizing Resource Utilization:** The use of AI systems significantly improves the efficiency of renewable energy projects [23]. The optimization of renewable energy installations requires engineering optimal farmlands for wind turbines and photovoltaics alongside precision energy output management and reductions in production waste.
- **Promoting Circular Economy Practices:** Circular Economy is envisaged through AI-driven analysis of the recycling and reutilization of elements of renewable energy systems. AI systems are then used to analyze component lifecycle data to identify ideal component reuse scenarios as well as materials recycling.
- **Monitoring and Mitigation of Environmental Risks:** Through monitoring services, AI detects environmental threats to wildlife and ecosystems and develops procedures that minimize these environmental risks. Through real-time data gathering and processing, biological systems receive proactive strategies that protect biodiversity.
- **Encouraging Renewable Energy Adoption:** Demonstrating low costs combined with improved performance of AI-integrated renewable energy systems will lead to increased adoption of sustainable energy systems [24]. The implementation and acceptance of initiatives benefit from public information through AI data-driven findings, which raise awareness and facilitate execution.
- **Long-Term Sustainability Metrics:** Through an AI system, researchers can design sustainable measurement frameworks to track renewable energy project sustainability across long periods. Sustainable renewable energy monitoring depends on three evaluation indicators that combine carbon intensity decline with operational efficiency achievements alongside community-based effects to foster progress across time.

FUTURE DIRECTIONS IN AI FOR WIND ENERGY

The wind energy sector benefits from advances in Artificial Intelligence as innovations emerge continually [25]. These technological advances improve the fundamentals of wind energy systems to enhance operational effectiveness and sustainability, alongside the development of scalable green energy solutions for transformative change.

EMERGING TRENDS IN AI TECHNOLOGIES

AI technologies are rapidly advancing, paving the way for novel applications in wind energy:

- **Advanced Machine Learning Models:** Sophisticated machine learning (ML) algorithms, including deep learning and reinforcement learning, improve prediction accuracy across wind speed measurements as well as power output and system performance behaviors.
- **Edge Computing:** Real-time data processing operations at wind turbine sites are now possible through the growing adoption of AI-powered edge devices. The technology minimizes system latency and eliminates reliance on cloud-based central locations, which leads to better operational efficiency.
- **Digital Twins:** Digital twin technology enables the development of virtual models both for individual wind turbines and complete wind farm complexes [26]. These digital duplicates demonstrate their value by delivering critical information on optimal performance and machine predictions, as well as failure tracking.
- **AI-Driven Control Systems:** Autonomous control systems demonstrate the power to tweak turbine procedures according to real-time weather patterns, together with power requirements.
- **Natural Language Processing (NLP):** Natural language processing tools now help organizations analyze maintenance reports, research papers, and operator feedback to support improved decision processes.
- **Integration of Quantum Computing:** Complex wind energy optimization tasks, such as layout design, energy flow management, and resource allocation, can be solved through quantum algorithms at levels unattainable by classical algorithms.

POTENTIAL FOR EXPANDING AI APPLICATIONS IN GREEN ENERGY

The role of AI in wind energy is set to expand, pushing for further applications in the green energy sector as a whole:

- **Hybrid Energy Systems:** The integration of Artificial Intelligence allows for the maximization of combined resource use and enhanced system reliability with renewable wind energy systems.
- **Energy Storage Optimization:** AI management systems optimize the storage of surplus wind power and control advanced high-capacity battery systems as well as pumped storage applications.
- **Grid Integration:** Artificial Intelligence can be introduced into smart grids for wind energy to promote its rapid integration in distributed power systems used for the spatial distribution of electricity for balance and stability of power grid operation [27].
- **Carbon Footprint Reduction:** The wind energy applications are carbon neutral due to their optimization of Artificial Intelligence, resulting in a loss of even 0.5% carbon footprint impact, monitored by the AI, for optimization of resources.
- **Microgrid Innovations:** The technology enables microgrids, *i.e.*, wind-derived energy systems, to address outlying areas that don't have access to power grids, further driving decentralized/sustainability-oriented energy development.
- **Climate Adaptation Strategies:** AI–based design functionality enables wind farm systems to evolve and adapt to climate shifts and changes in environmental conditions, providing reliable energy distribution across multiple systems.
- **Analysis Capabilities of AI Tools:** Such tools are able to analyze data about project impacts to help policymakers make informed decisions and develop supportive regulatory frameworks.
- **Global Collaboration Platforms:** Artificial Intelligence platforms allow global wind energy professionals to develop collaborative platforms to distribute operational data, research results, and share networks for industry best practices.

CONCLUSION

The latest progress on incorporating Artificial Intelligence in the wind energy sector signifies the future of power generation. AI's application solves critical issues of the wind power system with resource variability, as well as operational and high maintenance costs. The system provides precise wind resource evaluation and efficient power generation behavior through continuous monitoring systems powered by advanced predictive analytics and extremely affordable operational management. Beyond improving the efficiency of wind energy systems at the individual level, AI technology also plays a role in hybrid renewable power systems. We have developed an innovative class of advanced hybrid power systems that integrate hardware and software components to improve efficiency and stability, enhancing energy supply resilience and protective energy security functions. There are ethical and environmental implications to the deployment of AI, which ensure that integration is aligned with sustainability principles that foster just, sustainable development, which embodies environmentally responsible energy processes. As we edge closer to AI-based newer generation technologies such as Digital Twins, Edge computing, and Quantum Computing, the wind energy industry has been poised on the cusp of major developments. This kind of technological leap will not only open up new opportunities for Artificial Intelligence capabilities but will also significantly improve system scalability and integration with renewable energy grids. It also states that AI-based solutions are still evolving, so researchers should collaborate with policymakers and industry stakeholders to fully explore the potential of AI in renewable energy operations. The combination of Artificial Intelligence technology and wind energy is taking us exactly on the right path towards a sustainable and resilient green electricity future. AI thus represents a foundation to both create critical means to solve pervasive problems in our daily lives while finding ways to get us a sustainable present, and increasingly to find a better, sustainable future.

REFERENCES

[1] J. Ruan, Q. Wang, S. Chen, H. Lyu, G. Liang, J. Zhao, and Z.Y. Dong, "On vulnerability of renewable energy forecasting: adversarial learning attacks", *IEEE Trans. Industr. Inform.,* vol. 20, no. 3, pp. 3650-3663, 2024.
[http://dx.doi.org/10.1109/TII.2023.3313526]

[2] T. Qian, W. Ming, C. Shao, Q. Hu, X. Wang, J. Wu, and Z. Wu, "An edge intelligence-based framework for online scheduling of soft open points with energy storage", *IEEE Trans. Smart Grid,* vol. 15, no. 3, pp. 2934-2945, 2024.
[http://dx.doi.org/10.1109/TSG.2023.3330990]

[3] N. Sharma, M. Sharma, A. Singhal, R. Vyas, H. Malik, A. Afthanorhan, and M.A. Hossaini, "Recent trends in EEG-based motor imagery signal analysis and recognition: a comprehensive review", *IEEE Access,* vol. 11, pp. 80518-80542, 2023.
[http://dx.doi.org/10.1109/ACCESS.2023.3299497]

[4] M. Gordon, M. Daniel, A. Ajiboye, H. Uraiby, N.Y. Xu, R. Bartlett, J. Hanson, M. Haas, M. Spadafore, C. Grafton-Clarke, R.Y. Gasiea, C. Michie, J. Corral, B. Kwan, D. Dolmans, and S. Thammasitboon, "A scoping review of artificial intelligence in medical education: BEME Guide No. 84", *Med. Teach.,* vol. 46, no. 4, pp. 446-470, 2024.
[http://dx.doi.org/10.1080/0142159X.2024.2314198] [PMID: 38423127]

[5] D. Javeed, T. Gao, M.S. Saeed, and P. Kumar, "An intrusion detection system for edge-envisioned smart agriculture in extreme environment", *IEEE Internet Things J.,* vol. 11, no. 16, pp. 26866-26876, 2024.
[http://dx.doi.org/10.1109/JIOT.2023.3288544]

[6] I. Ul Haq, "Diagnosis of Neurological Disease Using Bioinspired Algorithms", In: *Bio-Inspired Optimization for Medical Data Mining,* S. Srivastava, A. Anand, A. Kumar, B. Saini, P.S. Rathore, Eds., Wiley-IEEE Press: Hoboken, NJ, USA, 2024, pp. 245-270.
[http://dx.doi.org/10.1002/9781394214211.ch12]

[7] M.B. Abdelghany, V. Mariani, D. Liuzza, O.R. Natale, and L. Glielmo, "A Unified Control Platform and Architecture for the Integration of Wind-Hydrogen Systems Into the Grid", *IEEE Trans. Autom. Sci. Eng.,* vol. 21, no. 3, pp. 4042-4057, 2024.

[http://dx.doi.org/10.1109/TASE.2023.3292029]

[8] D. Astolfi, F. De Caro, and A. Vaccaro, "Condition Monitoring of Wind Turbine Systems by Explainable Artificial Intelligence Techniques", *Sensors (Basel),* vol. 23, no. 12, p. 5376, 2023. [http://dx.doi.org/10.3390/s23125376] [PMID: 37420542]

[9] I.U. Haq, A. Kumar, and P.S. Rathore, "Machine learning approaches for wind power forecasting: a comprehensive review", *Discov. Appl. Sci.,* vol. 7, p. 1139, 2025. [http://dx.doi.org/10.1007/s42452-025-07675-x] [http://dx.doi.org/10.1002/9781394234271.ch10]

[10] P. Arora, S.M.J. Jalali, S. Ahmadian, B.K. Panigrahi, P.N. Suganthan, and A. Khosravi, "Probabilistic Wind Power Forecasting Using Optimized Deep Auto-Regressive Recurrent Neural Networks", *IEEE Trans. Industr. Inform.,* vol. 19, no. 3, pp. 2814-2825, 2023. [http://dx.doi.org/10.1109/TII.2022.3160696]

[11] A. Ahmadi, M. Talaei, M. Sadipour, A.M. Amani, and M. Jalili, "Deep Federated Learning-Based Privacy-Preserving Wind Power Forecasting", *IEEE Access,* vol. 11, pp. 39521-39530, 2023. [http://dx.doi.org/10.1109/ACCESS.2022.3232475]

[12] I. Ul Haq, G. Kaur, and A.H. Rather, "Fruit Leaf Classification Using Transfer Learning for Automation and Industrial Applications", In: *Deep Learning Techniques for Automation and Industrial Applications,* P.S. Rathore, S. Ahuja, S.R. Burri, A. Khunteta, A. Baliyan, A. Kumar, Eds., Wiley-IEEE Press: Hoboken, NJ, USA, 2024, pp. 115-138. [http://dx.doi.org/10.1002/9781394234271.ch10]

[13] F. K. Karim, D. S. Khafaga, M. M. Eid, S. K. Towfek, and H. K. Alkahtani, "A novel bio-inspired optimization algorithm design for wind power engineering applications time-series forecasting", *Biomimetics (Basel),* vol. 8, no. 3, p. 321, 2023. [http://dx.doi.org/10.3390/biomimetics8030321] [PMID: 37504209]

[14] F.M.A. Hadi, H.H. Aly, and T. Little, "Harmonics forecasting of wind and solar hybrid model driven by DFIG and PMSG Using ANN and ANFIS", *IEEE Access,* vol. 11, pp. 55413-55424, 2023. [http://dx.doi.org/10.1109/ACCESS.2023.3253047]

[15] I.U. Haq, and A.M.U.D. Khanday, "Exploring the role of explainable AI in women's health: challenges and solutions", In: *Explainable Artificial Intelligence in the Healthcare Industry,* A. Kumar, T.A. Kumar, P. Das, C. Sharma, A.K. Dubey, Eds., Wiley-IEEE Press: Hoboken, NJ, USA, 2025, pp. 283-306. [http://dx.doi.org/10.1002/9781394249312.ch13]

[16] A. Meng, S. Chen, Z. Ou, W. Ding, H. Zhou, J. Fan, and H. Yin, "A hybrid deep learning architecture for wind power prediction based on bi-attention mechanism and crisscross optimization", *Energy,* vol. 238, p. 121795, 2022. [http://dx.doi.org/10.1016/j.energy.2021.121795]

[17] C. Kaunert, B. Singh, R. Raghav, and A. Raghav, "Harnessing fuzzy logics in renewable wind energy for smart-sustainable cities and villages: Pioneering climate action through intelligent energy management in industry 5.0", In: *Fuzzy Logic in Revolutionizing Renewable Energy..* 1st ed., B. Singh, K. Vig, R. Lal, and B. A. Riswandi, Eds. New York, USA: Jenny Stanford Publishing, 2026, pp. 1–25.

[18] H. Dong, and X. Zhao, Wind-farm power tracking *via* preview-based robust reinforcement learning, *IEEE Trans. Industr. Inform.,* vol. 18, no. 3, pp. 1706-1715, 2022. [http://dx.doi.org/10.1109/TII.2021.3093300]

[19] F.K. Shaikh, M.A. Memon, N.A. Mahoto, S. Zeadally, and J. Nebhen, "Artificial intelligence best practices in smart agriculture", *IEEE Micro,* vol. 42, no. 1, pp. 17-24, 2022. [http://dx.doi.org/10.1109/MM.2021.3121279]

[20] S. Su, X. Yan, K. Agbossou, R. Chahine, and Y. Zong, "Artificial intelligence for hydrogen-based hybrid renewable energy systems: A review with case study", *J. Phys. Conf. Ser.,* vol. 2208, no. 1, p. 012013, 2022. [http://dx.doi.org/10.1088/1742-6596/2208/1/012013]

[21] M.S.H. Lipu, M.S. Miah, M.A. Hannan, A. Hussain, M.R. Sarker, A. Ayob, M.H.M. Saad, and M.S. Mahmud, "Artificial intelligence based hybrid forecasting approaches for wind power generation: progress, challenges and prospects", *IEEE Access,* vol. 9, pp. 102460-102489, 2021. [http://dx.doi.org/10.1109/ACCESS.2021.3097102]

[22] S. Higgins, and T. Stathopoulos, "Application of artificial intelligence to urban wind energy", *Build. Environ.,* vol. 197, p. 107848, 2021. [http://dx.doi.org/10.1016/j.buildenv.2021.107848]

[23] Y. Zhang, R. Li, and J. Zhang, "Optimization scheme of wind energy prediction based on artificial intelligence", *Environ. Sci. Pollut. Res. Int.,* vol. 28, no. 29, pp. 39966-39981, 2021. [http://dx.doi.org/10.1007/s11356-021-13516-2] [PMID: 33763837]

[24] L. Donadio, J. Fang, and F. Porté-Agel, "Numerical weather prediction and artificial neural network coupling for wind energy forecast", *Energies,* vol. 14, no. 2, p. 338, 2021. [http://dx.doi.org/10.3390/en14020338]

[25] P. Lu, L. Ye, Y. Zhao, B. Dai, M. Pei, and Y. Tang, "Review of meta-heuristic algorithms for wind power prediction: Methodologies, applications and challenges", *Appl. Energy,* vol. 301, p. 117446, 2021. [http://dx.doi.org/10.1016/j.apenergy.2021.117446]

[26] I. Ul Haq, I.U. Hassan, and H.A. Shah, "Machine Learning Techniques for Result Prediction of One Day International (ODI) Cricket Match", In: *Proceedings 2023 IEEE 8th Int. Conf. Converg. Technol. (I2CT)*, 2023. [http://dx.doi.org/10.1109/I2CT57861.2023.10126241]

[27] M.M. Rahman, M. Shakeri, S.K. Tiong, F. Khatun, N. Amin, J. Pasupuleti, and M.K. Hasan, "Prospective methodologies in hybrid renewable energy systems for energy prediction using artificial neural networks", *Sustainability (Basel),* vol. 13, no. 4, p. 2393, 2021. [http://dx.doi.org/10.3390/su13042393]

CHAPTER 2

AI-Driven Decision-Making Tools in Wind Energy

Vandita Nandal[1], **Himani Uppal**[1], **Inam Ul Haq**[2,*] and **Anand Kumar Gupta**[3]

[1] *Department of Computer Science and Engineering (CSE), Shree Guru Gobind Singh Tricentenary University (SGT) University, Gurugram, Haryana, India*

[2] *School of Engineering and Technology (SET), CGC University Mohali, Punjab, India*

[3] *School of Computer Science in Emerging Technologies, Noida Institute of Engineering and Technology (NIET), Greater Noida, Uttar Pradesh, India*

Abstract: The rapid evolution of Artificial Intelligence (AI) has opened new frontiers in optimizing wind energy systems, addressing long-standing challenges related to variability, inefficiency, and high operational costs. This chapter presents a comprehensive exploration of AI-driven decision-making tools applied throughout the wind energy lifecycle—from turbine design and site selection to forecasting, grid integration, predictive maintenance, and environmental protection. By leveraging machine learning, deep learning, and geospatial intelligence, AI enhances data analysis, improves accuracy, and supports real-time automation in energy management. Through real-world case studies from India, Iran, and Azerbaijan, the chapter highlights the practical benefits, scalability, and adaptability of AI across diverse energy landscapes. It also discusses the key advantages, emerging trends, and challenges associated with implementing AI in this domain. Ultimately, the chapter illustrates how AI is redefining wind energy development, contributing to smarter, more resilient, and sustainable energy systems worldwide.

Keywords: Artificial Intelligence (AI), Deep learning, Decision support systems, Energy optimization, Geospatial AI, Grid integration, Renewable energy, Site selection, Sustainable energy, Wind energy, Wind forecasting.

INTRODUCTION

A global shift toward sustainable energy solutions has made wind energy a necessary component of the renewable energy mix. There are many features in wind resources that are highly variable and complex, leading to significant problems in generation efficiency, system reliability, and minimizing operational costs. In response, using Artificial Intelligence (AI) in AI technologies, such as

* **Corresponding author Inam Ul Haq:** School of Engineering and Technology (SET), CGC University, Mohali, Punjab, India; E-mail: mirinam525@gmail.com

Sugandha Singh, Inam Ul Haq, Sachin Ahuja, Akib Mohi Ud Din Khanday & Neha Sharma (Eds.)

machine learning, deep learning, and expert systems, have demonstrated exceptional capabilities in processing large-scale, high-dimensional data typical of wind energy environments. These capabilities enable more accurate wind forecasting, efficient site selection, real-time turbine monitoring, predictive maintenance, and optimized grid integration. As a result, AI not only enhances operational efficiency but also contributes to the broader goals of reducing carbon emissions and achieving sustainable energy targets [1].

Decision complexities in wind energy systems are increasing with the larger wind farms being built, their linking with smart grids, and the need for real-time analytics. When AI-enabled tools are integrated into such systems, they will give adaptive learning and autonomous control features that enhance traditional rule-based methods [2]. Such tools will be able to dynamically evolve the wind energy sector into a revolutionary way to address such problems through intelligent decision-making systems.

Accommodating environmental uncertainties and changes in systems is expected to provide timely and reliable insights for the use of operators and planners. AI is also more than just technical optimization. It will help improve strategic arguments concerning the economic feasibility of wind projects, risk patterns, and investment strategies. AI's influence in this interdisciplinary manner is thus a pillar that supports intelligent energy infrastructure and hurries the entire world toward clean energy [3]. Increasingly, innovations in AI-powered robotics will improve maintenance methods relating to autonomous inspection and fault detection, leading to safer and shorter downtimes in remote and extreme environments [4]. All of these point to a trend toward increasing automation and resiliency in energy systems.

In essence, the confluence of AI technologies and wind energy serves more than just technological enhancement. It represents a shift in paradigm toward smarter, more sustainable, and future-ready energy solutions [5]. This chapter explores the multidimensional applications of AI in wind energy, assessing current tools, case studies, benefits, challenges, and future perspectives to provide a comprehensive understanding of this rapidly evolving domain.

ROLE OF AI IN WIND ENERGY

Artificial Intelligence is changing the way wind energy systems are heavily automated, intelligently driven in decision-making using data, and optimized at large scales. The industry's need for intelligent intervention by AI encompasses innovations across the board in turbine design, forecasting, site selection, environmental monitoring, and operations and maintenance, as demand surges along a mind-boggling cycle of variability in wind patterns and the need for sustainability. Table **1** shows applications of AI in wind energy and corresponding benefits across design, operations, forecasting, and sustainability.

Table 1. Applications of AI in wind energy and corresponding benefits.

Application Area	AI Techniques Used	Key Benefits
Wind Turbine Design Optimization	Genetic Algorithms (GA), Particle Swarm Optimization (PSO), Deep Learning (CNNs), Digital Twins	Optimizes blade design, improves aerodynamics, reduces prototyping costs, and enables real-time simulation and customization.
Wind Resource Assessment & Site Selection	Random Forests, Gradient Boosted Trees, Hybrid AI Systems, GIS with Fuzzy Logic & MCDM	Enhances wind mapping accuracy, identifies optimal sites, includes multi-criteria evaluation, and reduces siting risks.
Forecasting & Grid Integration	LSTM, GRU, Hybrid Forecasting Models, Reinforcement Learning	Increases prediction accuracy, improves grid stability, and enables smart dispatch and trading strategies.
Predictive Maintenance	Neural Networks, Support Vector Machines (SVM), K-Means, Autoencoders, Drone Vision Models	Detects faults early, reduces downtime and maintenance costs, and improves safety and reliability.
Environmental Protection	Computer Vision, Radar Fusion, Wildlife Migration Models, Acoustic Data Neural Networks	Prevents bird-bat collisions, minimizes ecological impacts, and models noise patterns for community acceptance.

Wind Turbine Design Optimization

The use of AI technologies has transformed wind turbine design from an inherently hands-on and trial-and-error process to the one built on a highly data-driven iterative design optimization path. With advanced machine-learning algorithms, engineers can simulate, evaluate, and optimize thousands of design options for turbine components in a small fraction of the time required by classical computational fluid dynamics (CFD) models.

- Genetic algorithms (GA) and particle swarm optimization (PSO) have gained wide acceptance for optimizing blade aerodynamics, hub designs, and material usage to improve performance and cope with structural resilience [6 - 9]. These algorithms explore large multidimensional design spaces and converge towards high-efficiency designs while mimicking natural evolutionary processes.
- Deep learning, particularly utilizing convolutional neural networks, is being coupled with aerodynamic modeling software to predict flow behavior and pressure distributions over turbine blades. Nasir *et al.* [10] showed that AI systems can now learn blade performance characteristics over varying wind speeds and turbulence levels, allowing improved turbine customization to site-

specific wind regimes.

- Increasingly, digital twins—virtual replicas of physical turbines—are used alongside AI to run real-time simulations. The digital twin receives real-time sensor data and dynamically adjusts design parameters to optimize energy capture for the present conditions [11].

By minimizing physical prototyping requirements, AI allows manufacturers to lower development costs and accelerate the time-to-market of high-efficiency turbines.

Wind Resource Assessment and Site Selection

The site selection requires a deeper understanding of wind behavior over time and space, local terrain particulars, environmental specifications, and grid system accessibility. Traditional models are poor at granularity; however, Artificial Intelligence has shifted site assessment into a new dimension by analysing large, heterogeneous datasets.

- Machine learning algorithms like Random Forests and Gradient Boosted Trees are used to process meteorological data, satellite images, and topographic maps to generate high-resolution wind resource assessments [12], and predictions of wind speed, direction, and variability local to micro-locations with greater fidelity than traditional mesoscale approaches.
- Thus, supervised and unsupervised hybrid AI systems allow the evaluation of sites continuously updated with changes in real-time weather and climatic conditions. In their work, Simankov *et al.* [12] highlighted the advantage of hybrid systems in securing the long-term sustainability and performance of wind farms amid changes in wind regimes and environmental conditions.
- GIS-based AI tools are also starting to get traction. When combined with fuzzy logic and MCDM frameworks, these systems enable developers to consider multiple siting criteria simultaneously, such as noise, wildlife habitats, road access, and proximity to transmission lines [13].
- Moreover, AI models predict greater turbulence and wake effects from buildings in urban settings than traditional siting models do. Higgins and Stathopoulos [14] have demonstrated the benefits of the urban wind energy project and AI-powered flow simulation for airflow prediction along flow paths within complex architectural environments.

With AI's intervention, recognition of previously overlooked high-potential sites is now possible while minimizing environmental and social risks associated with poorly chosen locations.

Forecasting and Grid Integration

One of the most prominent capabilities of AI in wind energy is energy output forecasting, which is vital for balancing real-time demand with supply in marketplaces. Wind is dynamic and uncertain by nature, causing difficulty in grid stability and the quality of power. AI models have a feature that addresses this uncertainty using predictive capabilities with high accuracy.

- Deep learning networks such as Long Short-Term Memory (LSTM) and Gated Recurrent Units (GRUs) for temporal dependencies modeling in wind speed and direction measurement data, short-, medium-, and long-term forecasting, respectively [15]. The models were reported to be superior to the typical statistical methods when it comes to sudden changes in wind behavior prediction.
- AI-based hybrid forecasting models constitute a combination of physical atmospheric models blended with data-driven machine learning to comprise resilient systems of predictions across the climatic zones. According to Chatterjee and Dethlefs [16], indeed, utility companies now employ these hybrid models to make real-time load balancing and grid stabilization decisions.
- Grid management tools are now under AI and able to forecast power output from many wind farms at a time, adding to the forecasting model inputs weather forecasts and turbine conditions data, as well as market demand profiles. It could also adjust energy storage and dispatch in real time to achieve reliability for the grid.
- AI in Energy Trading: High-sophistication reinforcement learning techniques thus enable power producers to take on better roles concerning activity in power markets by learning about optimal bidding strategies corresponding to changes in price and even expected generation in the market [17].

Increased reliability and scalability in national and regional energy portfolios for wind power are achieved by making forecasts more accurate, with the possibility of seamless integration into the grid.

PREDICTIVE MAINTENANCE AND ENVIRONMENTAL PROTECTION

Maintenance costs are an important operational cost since wind turbines are exposed to harsh environmental conditions and mechanical stress. Modern time-based maintenance methods are not efficient and often lead to unplanned downtimes. The AI approach to maintenance introduces an anticipatory paradigm: predictive maintenance, which can considerably reduce operational costs.

- The supervised learning models are trained on SCADA (supervisory control and data acquisition) data to study early signs of mechanical deterioration, such as increased vibration, excessive heating, or unusual pitch control activity. Garcia Marquez and Peinado Gonzalo [17] showed that fault occurrence could be predicted by neural networks and support vector machines (SVM) with an accuracy of over 90%.
- The unsupervised methods include k-means clustering and autoencoders for anomaly detection in historical performances, thereby helping isolate small deviations from normal operating behavior, from which escalation can be avoided.
- AI-powered robotics and drone technology are now playing an increasing role in wind turbine blade inspection, where real-time image recognition models detect cracks, erosion, and lightning damage [18]. This also helps to minimize human intervention and enhance safety.
- From the green point of view, AI helps mitigate ecological effects resulting from wind turbines. Collision avoidance systems use computer vision, fused with radar data, to follow the movements of birds and bats near turbines and automatically intervene to make them halt blade rotation during identified high-risk outs [9].
- With the aid of wildlife migration models, powered by AI, wind farm operators can adjust the operational schedules biannually for biodiversity conservation. Khan and Nasir [19] have shown how the satellite and acoustic data integrated with neural networks allow tracking of migratory bird patterns.
- Additional applications of AI on social and environmental noise emitted by turbines are being investigated. By modeling wind speed, direction, and turbine positioning, predictive models estimate noise impact areas and suggest appropriate design changes or operational time adjustments.

AI not only makes wind farms smarter and more effective but also brings about ecological sustainability and acceptance in the local communities and surrounding ecosystems.

AI APPLICATIONS IN WIND ENERGY

The way data is interpreted, decisions are made, and resources are optimized is being transformed by Artificial Intelligence (AI) in wind energy systems. Fig. (**1**) shows that Applications of Artificial Intelligence in Wind Energy Systems encompass the entire lifecycle of a wind energy project: planning and deployment, real-time operations, and maintenance. AI tools try to enhance forecasting accuracy, optimize sites, monitor turbine health, and forecast responses to potential failures by taking advantage of vast datasets and advanced computational models. They aim at improving overall efficiency and reducing operational risks while developing smarter and more sustainable infrastructures for energy.

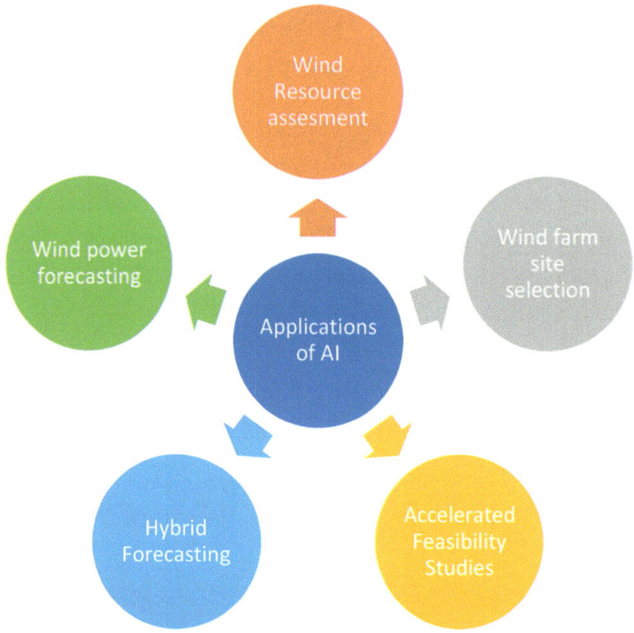

Fig. (1). Key applications of artificial intelligence in wind energy systems.

Wind Resource Assessment and Site Selection

Wind resource assessment and site selection are primary activities in wind energy development, deciding the longevity of the project performance and economic viability. With the advent of artificial intelligence, especially machine learning (ML) and geospatial analytics, developers can make accurate and data-driven project decisions both quickly and confidently. The technology helps analyze large-scale, multi-dimensional datasets that cover parameters of historical weather patterns, topography elevation features, and environmental constraints, allowing for improved site feasibility, reduced project risk, and accelerated path of wind farm deployment.

- **Use of machine learning algorithms (Random Forest, SVM) for analyzing historical weather, topography, and wind data:** Machine learning algorithms such as Random Forest (RF) and Support Vector Machines (SVM) have proven to be highly effective in analyzing historical meteorological parameters such as wind speed, temperature, and atmospheric pressure. The advantage of such models is their capability of detecting nonlinearities in huge data sets, which makes them very appropriate for accuracy in modeling wind behavior under terrain conditions. For instance, the RF model has been shown to predict wind speeds based on past weather observations combined with terrain features, producing reliable results better than those obtained with traditional statistical methods [20]. In principle, the ML schemes are integrated with output from numerical weather prediction (NWP) for bias correction and better surface wind speed prediction, especially in complex areas with fewer measurements [21].

- **Geospatial AI models to optimize the location of wind farms for maximum energy yield and minimal environmental impact:** Geospatial AI involves the incorporation of AI models with GIS (Geographic Information Systems), satellite imagery, and digital elevation models for wind farm site identification. All these systems assess many variables such as land use patterns, slope, elevation, proximity to infrastructure, and environmental restrictions to specify energy-friendly but eco-socially considerate sites. The whole approach is a multi-criteria decision system for broader and sustainable planning, as AI-assisted geospatial tools ensure spatial clustering and zoning optimization; these reduce very wide areas into specific high-potential zones with much more precision and less manual effort [21].

- **AI reduces time and uncertainty in traditional feasibility studies:** It is often time-consuming and uncertain for a classic feasibility study on a site, requiring an extended data collection from a larger number of field measurements. Artificial intelligence will dramatically accelerate those activities by automating data integration, analysis, and modeling of scenarios. For example, historical climate-topographic datasets applied through machine learning techniques can result in an AI system that can mimic energy outputs, assessing site viability with one of the varying climatic conditions [22]. Those predictive capabilities offer faster, evidence-based insights to stakeholders and reduce dependence on cumbersome on site studies, thus shortening the development period and investment risks.

Wind Power Forecasting

Accurate wind power forecasting plays a pivotal role in the operational efficiency and financial viability of wind energy systems. Due to the inherent variability and intermittency of wind, real-time and ahead-of-time forecasting is essential for maintaining grid stability, scheduling power generation, and reducing reliance on fossil fuel-based reserves. Traditional forecasting methods often fall short in capturing the nonlinear and chaotic behavior of wind. However, the advent of deep learning and hybrid modeling techniques has significantly enhanced the accuracy and reliability of wind forecasts. By combining physical simulations, statistical models, and advanced AI algorithms, modern forecasting tools support more adaptive and intelligent grid operations.

- **Use of deep learning models (like LSTM, CNNs) to predict short-term and long-term wind speeds and energy output:** The deep learning models of LSTM and CNN, in particular, perform well with the ability to capture temporal and spatial features of wind data. Because of their ability to address nonlinearities, seasonality, and sudden fluctuations associated with wind behavior, the mentioned models are quite effective. As an example, convolutional LSTM networks, when joined with wavelet packet decomposition, significantly enhance the accuracy of both short-term and long-term wind speed forecasts [23]. These models provide highly granular real-time decision-making inputs to energy production by learning from historical time-series data.

- **Hybrid models combining physical and statistical forecasting approaches:** Hybrid forecasting models embody a powerful blend of domain knowledge with data-oriented insight. These models link dynamical (physics-based) simulations with machine learning algorithms to leverage the best of both worlds. The physical models maintain fundamental understanding through fluid dynamics and atmospheric physics, while the AI methods enhance predictions through bias correction and learning from residual patterns based on historical errors. For example, hybrid approaches that combine statistical methods with ML routes, such as ensemble regression, support vector regression, or neural networks, provide considerable gain in prediction accuracy, especially under uncertainty or rapidly changing weather situations [24]. Thereby helping to connect theoretical modeling with practical forecasting.

- **Improves scheduling, load balancing, and reduces reliance on fossil fuel backups:** By improving the matching of supply with actual demand, accurate forecasting will greatly assist operators on the grid in achieving overall load balancing and operational planning. With confident wind prediction, energy dispatching schedules would then improve to curtail excesses and reduce economic losses further. Quality forecasts would reduce dependence on fossil-fueled peaking plants that were customarily used as backups during wind lulls. Hence, AI-enhanced forecasting directly promotes more stable and sustainable grid operations, aiding the larger transition to low-carbon, locally balanced energy systems.

CASE STUDIES

AI-Driven Innovations in Wind Energy Projects in India

As one of the fastest-growing renewable energy markets in the world, India is now using Artificial Intelligence (AI) to eliminate operational challenges in wind energy generation and to achieve its clean energy targets. This study discusses various AI real-world applications in wind energy projects in India, illustrating

the extent to which decision-making, maintenance planning, and energy optimization are carried out using advanced data analytics and intelligent systems.

AI has found numerous implementations for value addition and has significant areas of load forecasting. Energy forecasts were largely statistical models that were not completely reliable, given massive wind fluctuations. AI models have now found useful applications,the AI models typically use supervised learning algorithms, in forecasting in Indian wind farms based on massive data sets, which include historical weather conditions, wind speed, and energy demand cycles, among others. Mostly based on Long Short-Term Memory (LSTM) networks and random forest algorithms, these models can now predict wind energy generation fluctuations with very high accuracy. This is especially important for grid stability in regions like Tamil Nadu and Gujarat, as irregular wind generation puts pressure on the grids. A well-coordinated forecast will enhance the grid reliability of these regions and reduce further dependency on fossil fuel-based generation when coordinated with storage and other thermal backups.

Another exciting novelty is that of predictive maintenance strategies using AI. Sensor data on parameter values such as rotor speed, vibration frequency, bearing temperature, and acoustic signatures is collected in real-time by turbine sensors at many Indian wind farms. AI models based on data from historical failures can provide early warning signs of mechanical degradation or malfunctioning components. Instead of conducting maintenance activities under unplanned downtimes, maintenance can be planned proactively to save costs in maintenance and downtime, thereby boosting turbine reliability and energy output. AI-enabled condition monitoring systems have been deployed in wind farms run by both Suzlon and ReNew Power, leading to a significant increase in turbine availability and overall energy generation from these projects.

The case study also describes how the optimization is done in operational procedures and resource allocations through AI. The most efficient operational settings for turbines based on the current wind conditions and market demands are determined through the utilization of optimization algorithms like genetic algorithms and particle swarm optimization. These algorithms also play an important role in dispatch planning by aiding the energy suppliers in the selection of the best combination of turbines that optimizes output while minimizing operational stress on the turbine components.

In the area of strategic planning, geospatial AI models are used for site selection of new wind farms. Armed with satellite data, GIS layers, land use maps, and other economic factors, these models search for optimal locations for development. In India, where land availability and environmental impacts are key driving factors, AI has streamlined feasibility studies and helped cut more than 40% of the time needed for site evaluation, according to project reports cited within the paper.

AI platforms are also enhancing energy management systems (EMS) in integrated renewable energy parks, whereby these platforms integrate the data from the wind, solar, and storage systems to balance the flows of energy. In the hybrid parks under the Gujarat Hybrid Renewable Energy Park initiative, for instance, AI algorithms manage real-time dispatch and storage decisions so that energy delivery can remain intact even with erratic weather conditions.

Last but not least, the case study appraises socio-economic and environmental benefits from these AI innovations. An improved predictive maintenance system means fewer emissions and less waste, leading to a more stable and cheaper energy supply. AI-powered microgrid systems powered by wind energy have great potential to aid in local electrification and economic upliftment in rural and far-off areas with limited access to the grid.

This case study from India demonstrates the transformational prospect of AI that could lend modernization to wind energy systems. It showcases not only technological aspects but also that the AI deployed complies with national energy policies and sustainable development goals. The integration of AI in India is emerging as the flagship case for other developing countries that want scalable, efficient, and clean energy solutions [25].

Wind Farm Site Selection in Semnan Province, Iran

The integration of Artificial Intelligence with the GIS model has swung the wind farm site selection process away from manual, labor-intensive undertakings and into a highly systematic and precise realm of decision-making. This case study, focusing on Semnan Province in Iran, compared applications of the AI-enabled MCDM methods with geospatial analytics to identify sites for wind energy development.

The proposition of northern Iran as an area with an interesting but intricate blend of wind conditions and geographical features makes Semnan Province attractive yet challenging for wind energy deployment. Therefore, the study by Yousefi *et al.* (2022) presented a highly sophisticated framework that unified GIS-based spatial analysis with an MCDM approach powered by AI algorithms for concurrently addressing multiple layers of environmental, technical, and infrastructural criteria [26].

The manuscript's most relevant variables under the AI-GIS framework included average wind speed, topography (elevation and slope), land use classification, and distances to power lines and the roadway networks, as well as environmental constraints in terms of protected areas and urban settings. The layers were run under the Analytic Hierarchy Process (AHP), whereby a structured MCDM approach supported by AI-based weighting algorithms ensured that the level of importance assigned to each criterion reflected the underlying knowledge and information derived from the empirical data.

Wind speed data, the most vital parameter in the evaluation process, were derived from long-term meteorological records and satellite-based models. AI models were then applied to interpolate the wind speed estimates and add more spatial resolution to the estimates. This step significantly heightened the accuracy in setting wind resource mapping and distinguishing areas at micro-locations with high energy generation potential. Remote sensing techniques were applied to land-surface data, while AI was employed to classify terrain and land cover with superior accuracy to distinguish between areas that were technically suitable versus those that were legally or environmentally constrained.

The main output of the AI-GIS unification was a suitability map, clearly identifying areas that were categorized as high, moderate, and low suitability for wind farm development. Energy generation and infrastructural feasibility requirements were met in various clusters highlighted in the study within Semnan Province. These areas were located near existing power transmission lines and road accessibility; thus, potential project costs related to construction and grid integration would be lower.

One of the important contributions that this case study can boast of is reducing the uncertainty and delay in decisions concerning large-scale renewable energy planning. Conventional site selection methods cannot negotiate the different conflicting factors. Yet, an MCDM model with embedded AI showed that a data-driven system could juxtapose technical feasibility with ecological preservation and socio-economic aspects. Moreover, the framework developed herein is scalable and capable of being tuned to meet the needs of other provinces or countries that face similar challenges in the sort of geography and energy planning.

The authors further indicated that the adoption of this approach facilitates transparent and replicable decision-making, giving stakeholders, from government planners and investors to environmental regulators, a lucid and evidence-based direction to follow in selecting specific sites. Such AI-based systems are going to be an important basis for cost-efficient and environmentally sustainable wind energy expansion as Iran seeks to diversify its energy portfolio in the international context.

As a summary, the case study of Semnan is an interesting example of how the combination of AI and GIS technologies with developed decision-making models can change the landscapes across which strategic planning of wind energy facility development takes place. The practical advantages of incorporating high-resolution spatial information with AI logic to minimize the time frame of renewable energy deployment and its associated uncertainties have emerged from this case [27].

Machine Learning for Wind Farm Operations in Azerbaijan

In moving into predictive maintenance strategies, wind generation network operations in Azerbaijan are now strategically deploying machine learning (ML) models amid a rapidly expanding wind infrastructure where investments keep pouring into renewable energy. As Mammadov *et al.* (2024) show, the more developed countries have already entered into advanced data-driven tools to optimize not only operational performance gain but also energy output forecasting and adaptive control of wind turbines under varying weather conditions [28].

The core concern of the study is how machine learning algorithms, such as Decision Trees (DT), Support Vector Machines (SVM), and Artificial Neural Networks (ANN), have been used to improve the predictive capability of a wind farm system. The data set for the training consists of historical meteorological data, turbine performance data, and real-time response inputs from sensors. The ultimate goal was to improve short-term and mid-term predictions of electricity generation that would enable operators to make the most assured scheduling and dispatch planning.

And yet, one of the fundamental challenges to be addressed in the management of wind energy is the intermittency inherent in, and variability of, wind resources. The study demonstrated that machine learning methods perform better than pure physical models since they build innumerable complex nonlinear laws describing the wind behavior and turbine response. For example, an ANN trained the systems to remember previous patterns and modify the forecasts dynamically when exposed to weather anomalies or turbine defects that had not been previously encountered.

One of the breakthroughs tested in the Azerbaijani wind farms was the use of AI-powered adaptive control techniques. These systems used immediate feedback to change the angles of turbine blades, yaw directions, and load conditions of generators. Processing real-time data through ML models, the turbines could change under varying speeds of wind, reducing energy losses and enhancing efficiency overall. Such intelligent automation considerably decreased lagging in operations and dependency on human interventions, which are usually more costly and error-prone.

Beyond this, the research delineated predictive maintenance opportunities brought about by machine learning deployment. Wind park operators could trigger early warning indicators into the mechanism of mechanical wear or parts failure by using DT and SVM trained on historical failure data and real-time measurements of turbine sensors, such as those measuring vibration, temperature, and acoustic emissions. The operators would halt operations to conduct maintenance on different occasions of low demand, minimizing time lost in production and avoiding costly breakdowns.

It also emerges from the case study that AI has a wider impact on grid reliability and market participation in energy due to the improvement in forecast accuracy. Wind-generated energy improves the synchrony between what is offered to the grid and the demand on it, creating a much more stable supply of energy. This was of interest to Azerbaijan because there are more plans put forward by wind farms in this country to cater to national energy targets and cut down dependence on fossil fuels.

The study also stressed that these AI solutions can be scaled and made adaptable. The AI tools developed for wind farms in Azerbaijan are modular, meaning retraining and customization can be undertaken for other regions with different climates and topography. It is an indication of the potential for spreading AI-based responses across other developing countries wanting to modernize their renewable energy systems.

Thus, the case study concerning Azerbaijan indicates that machine learning not only enhances the operational efficiency of wind farms. Rather, it also promotes the development of intelligent energy management systems. Predictive analytics, adaptive control, and intelligent maintenance are such areas in which AI finds an important role to play in modernizing wind power operations and making renewable energy systems more resilient and more data-informed [29].

AI in Operations and Maintenance (O&M) of Wind Turbines

This case study discusses in detail how AI is bringing a paradigm shift in the Operations and Maintenance (O&M) environment of wind turbines. In their comprehensive scientometric analysis, Chatterjee and Dethlefs (2021) present data showcasing an increasing use of AI-assisted models—mainly Support Vector

Machines (SVMs), Artificial Neural Networks (ANNs), and Deep Learning architectures—for condition monitoring and predictive maintenance in real time.

Until recently, traditional maintenance systems tried to address wind turbine maintenance further by having the operator rely on either fixed inspection schedules or reactive servicing, which, in general, caused delays in fault detection and mediation with higher operational costs. Through this study, it is highlighted that AI models build this process so much better with predictive names, whereby faults can be diagnosed early before the critical actual failure.

The authors discuss several instance cases wherein SVMs and deep learning models have been applied by using large datasets generated by turbine sensors, parameters including vibration signals, temperature readings, acoustic emissions, and generator torque data. These AI models distinguished normal operational behavior from early signs of mechanical anomalies such as gear wear, bearing failure, and rotor imbalance with accuracy.

One of the prominent benefits of using AI in O&M is its ability to detect nonlinear and hidden patterns within massive streams of operational data. For example, deep learning models, particularly Convolutional Neural Networks (CNNs), can identify minute changes within sensor signals that are usually beyond the reach of conventional monitoring techniques. Plugged into turbine control, these models automatically trigger preventive actions, thus drastically reducing unplanned downtimes.

The case study also offers clear examples of cost benefits and increased availability, driven by AI maintenance methods. Wind operators have reported better uptime of turbines and fewer frequent calls for maintenance since actions were only initiated as necessary, which preserved not only the health of critically located components but also enhanced the useful lifetime of the turbine infrastructure, thus maximizing the return on investment made from renewable energy projects.

A key finding of the study showed an emerging trend towards autonomous O&M systems whereby AI techniques are integrated into the IoT networks on the turbines, providing them with the means for remote diagnostics and real-time performance evaluation, thus reducing incidences of manual inspections, especially offshore or in hard-to-reach areas.

The review also highlights barriers such as the need for high-quality, labeled training data and the interpretability of AI decisions when applied to high-risk industrial environments. Nevertheless, advances in explainable AI (XAI) and

federated learning will probably result in an increase in robustness and transparency of such systems in the coming times.

In summary, the case study shows how AI has transformed O&M in the wind energy sector: from the old pattern of reactive maintenance to modern intelligent, predictive, and autonomous servicing models. These changes are not only contributing to great reliability in operations but will further aid in energy sustainability and cost-effective management of infrastructure [30]. Table **2** summarizes key global case studies that highlight the impact of emerging technologies on sustainable resource management.

Table 2. Summary of global case studies.

Case Study	Focus Area	Key AI Techniques & Outcomes
India – AI in Wind Energy Projects	Forecasting, Maintenance, EMS, Site Selection	LSTM, RF, GA, PSO → Improved forecasting, predictive maintenance, optimized planning
Iran – Wind Farm Site Selection	GIS-based Site Selection & Feasibility Analysis	AI-AHP, Remote Sensing → Accurate wind mapping, suitability zones, faster studies
Azerbaijan – ML in Wind Farm Operations	Forecasting, Adaptive Control, Predictive Maintenance	DT, SVM, ANN → Real-time control, better forecasting, grid stability
Global – AI in Wind Turbine O&M	Operations & Maintenance	SVM, CNN, IoT → Fault detection, autonomous systems, longer turbine life, fewer manual checks

BENEFITS OF AI-DRIVEN TOOLS

AI technology has brought a revolutionary change in the wind energy industry, and it is beyond the planning or construction phase; in fact, it extends into the lifetime of the wind farms with performance optimization. Such tools allow for better decision-making, improvement of operational efficiencies, and assist in sustainability both at an environmental and economic scale. Below are some salient advantages of AI-empowered systems in wind energy applications.

Greater Accuracy and Speed in Code Review Process: AI-enabled automatic code review instruments have revolutionized the process of software development by instantly identifying programming mistakes, potential bugs, and security threats. Consequently, these systems operate with high accuracy and consistency while taking up little time for manual reviews. Therefore, they allow for quick feedback cycles and continuous integration in agile environments, making the resulting codebase of higher quality, as well as the development workflow more effective [31].

Enhanced Creativity in Design and Art: AI tools act as co-creative entities in the creative industry, assisting with suggestions, conceptualizing, or even filling in the blanks in a visual masterpiece. Artists and designers are using these systems to come up with quick abstractions and complex possibilities in design, allowing faster iteration and sparking creativity in areas like game design, digital illustration, and animation, thus freeing these professionals from repetitive tasks and making them more productive at creating masterpiece designs [31].

Increased Efficacy in Performance Testing: In performance testing, AI technologies such as machine learning help in generating smarter test scenarios and automatically adapting tests depending on real-time application response validation and coverage of all performance issues without the need for any manual intervention. These tools can display latency problems, identify server bottlenecks, and highlight system stress points faster than traditional means. The outcome will be a much more robust and scalable application needing less time and effort in carrying out the test process [32].

Enhanced Awareness of AI Limitations and Governance Requirements: Understanding what AI more effectively carries and cannot do is a vital foundation to derive the best out of the tool in both industries and research. AI processes vary, but usually introduce some attack forms, such as data dependency, algorithm bias, or overfitting, to name a few. To such an organization, fulfilling or not fulfilling such would mean having a weight on the positive part of AI and being saved or at least protected by limitations of governance, transparency, and human oversight [26].

Competitive Edge Through Digital Transformation: Much of the competitive improvement that AI has brought to digital transformation companies concerns customer targeting, supply chain efficiency, and improved decision-making. Personalization in marketing campaigns, sales forecasting, and automation of business processes form part of the AI tools. For example, LPP was able to use AI to improve understanding of customer habits and enhance inventory planning. Thus, these innovations provide competitive advantages to companies operating in very rapidly evolving markets. As illustrated in Fig. (**2**), AI-driven tools offer numerous benefits such as enhanced predictive accuracy, real-time monitoring, and optimized decision-making.

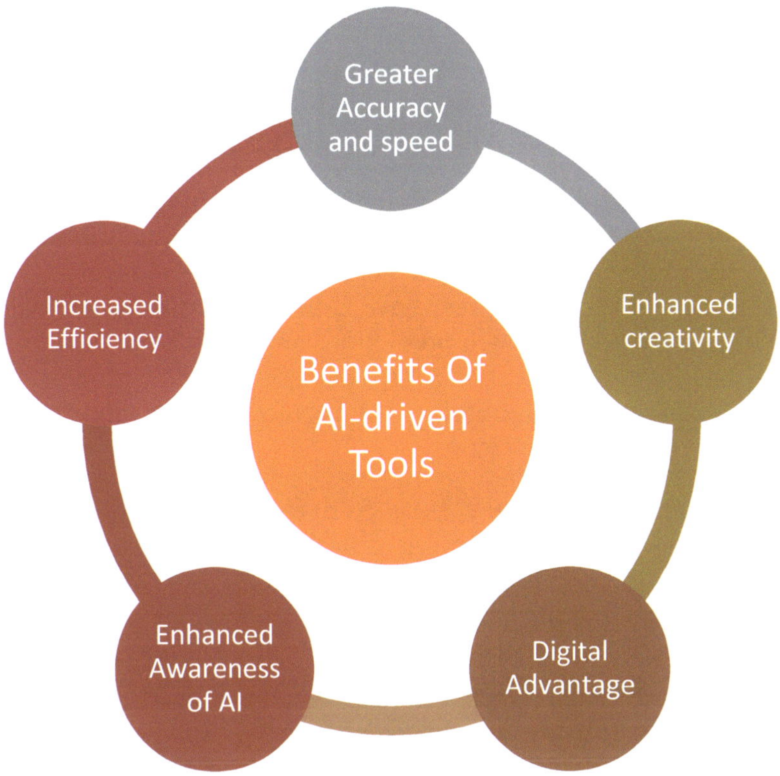

Fig. (2). Benefits of AI-driven tools.

CHALLENGES AND LIMITATIONS

Despite its transformative potential, the integration of AI in wind energy faces numerous roadblocks. From technical limitations to regulatory constraints, these challenges hinder seamless adoption. A critical examination of these issues helps uncover gaps in implementation. Addressing them is key to realizing AI's full benefits in the renewable energy domain.

Challenges in AI-Driven Decision Making for Wind Energy

Though a formidable game changer, the deployment of artificial intelligence (AI) in the wind energy sector is not without significant barriers. Such barriers include technical, socio-economic, and policy-related barriers, all of which hinder the widespread adoption of AI-based decision support systems. Understanding these challenges is vital in enhancing the reliability, scalability, and acceptance of AI tools in wind energy planning and operations.

Complexity of Spatial Planning: Several spatial, environmental, and socio-political aspects determine the placement of wind turbines, thus rendering the outcome of such decisions complex in the decision-making process. The integration of these criteria into AI models represents a major challenge, even when using spatial decision support systems [33]. Conflicts occurring in land uses, ecological sensitivity, or zoning requirements further complicate the situation and make the use of AI very specific to context.

Data Availability and Quality: Efficient and effective application of AI is coupled with the need for enormous, accurate, and high-resolution data volumes for most of the AI applications. The inadequacies in most areas, especially offshore or in developing countries, have limited these kinds of data, thereby affecting the reliability upon which predictions and optimizations rely [34]. Also, scarcity of data tends to decrease the performance of models or lead to misinformed decisions, thus reiterating the requirement of robust infrastructure for data collection and standardization.

Integration of Multidisciplinary Data Sources: The integration of data and information from meteorological, geographical, mechanical, and market sources into one AI framework is complicated. Such complexity is compounded in real-time applications that require simultaneous processing of data generated from sensors, weather feeds, and grid systems. Synchronization and harmonization of such varied data streams require sophisticated preprocessing and fusion techniques, which increase system complexity and cost.

Forecasting Accuracy and Variability: The stochasticity of wind also makes it difficult to generate predictions that can maintain accuracy over the important intervals, although AI has been making strides in its forecasting abilities. Currently, short-term fluctuations continue to threaten the schedules and stability of grids [35]. As a result of these inaccuracies, there is almost certain imbalance in energy supply, increased reliance on backup systems, and less confidence among grid operators.

Limitations of AI-Driven Decision-Making for Wind Energy

In optimizing wind energy processes, AI-driven decision-making tools prove essential; however, their inherent limitations would have to be dealt with if they are to gain more acceptance. Such limitations range from technical to operational and economic factors impeding AI's fullest potential in wind energy. Perspective of such limitations enables the countermeasure development and subsequent fine-tuning of tools for a better fit in practical applications in the energy sector.

Limited Generalizability Across Geographies: Datasets used to train AI models sometimes originate from very specific regions that may generalize poorly to other geographies with different wind patterns, topographies, or other environmental conditions. The licenses of those models that were trained in a given location could be compromised in a region with much different characteristics. Thus, this limitation opens a call for a region-specific customization or even retraining of the models to allow accounting of local variations in terms of wind resource availability and infrastructure.

Dependence on Historical Data for Training: A considerable amount of historical data plays a significant role in developing AI models, in particular, machine-learning algorithms, and is essentially required for proper training. The quality of predictions is greatly dependent on the amount and heterogeneity of historical data that are used. For instance, in an area endowed with minimal or inconsistent amounts of historical data, any AI-driven prediction will lose some degree of accuracy and reliability. This situation creates the challenge of having either too little data, thus yielding overfitting, or, at best, compromising performance, which could lead to inefficiencies or higher operational risks in making real-time operational judgments.

Lack of Real-Time Adaptability: If these tools can truly enhance II prediction, then their capacity for real-time adaptation to sudden environmental changes (shifting wind speed, for example, or one-off extreme weather events) remains limited. AI models require continuous updates and recalibration; however, ensuring that this is done in dynamic conditions that change every minute is difficult. Such limitations inhibit AI's ability to manage unforeseen operational changes or emergencies, hence weakening the role it plays in determining critical decisions.

High Computational and Energy Requirements: AI models, especially deep learning-based and those with intensive algorithms, need computational power for training purposes as well as for real-time inferencing. Hence, it demands high expenditure for AI deployment in wind farms that are remote or resource-constrained [36]. Such operational overhead substantially increases the cost-effectiveness of AI solutions, primarily for small-scale or off-grid wind projects wherein resources are limited.

Ethical and Social Concerns: AI systems are immensely powerful yet operate in a black box about transparency in decision-making, and consequently, the question of accountability and ethics comes into play. Concerning wind energy, AI systems, when making decisions for sites or environmental impact assessment, may neglect community or ecological concerns, eventually leading to public resistance or legal challenges [37]. Therefore, lies a demand for a more inclusive decision-making process that involves stakeholders and guarantees AI recommendations do not diverge from societal and environmental considerations.

CONCLUSION

The integration of Artificial Intelligence into the wind energy sector has marked a transformative shift in how modern energy systems are designed, managed, and optimized. From improving wind turbine design and site selection to enabling accurate forecasting and predictive maintenance, AI has emerged as a critical enabler of efficiency and sustainability. Through advanced algorithms, real-time analytics, and intelligent automation, AI enhances decision-making across all phases of wind energy development, ultimately contributing to reduced operational costs, improved grid stability, and lower environmental impact.

The case studies explored in this chapter—from India's large-scale AI-enabled forecasting systems to Iran's geospatial optimization models—highlight the global potential and adaptability of AI across diverse geographies and operational challenges. Despite its promise, the technology's adoption is not without hurdles, including data limitations, integration complexities, and the need for regulatory alignment. However, with continuous advancements in machine learning, sensor technology, and explainable AI, many of these challenges are expected to be mitigated over time.

In conclusion, AI is not just enhancing the technological backbone of wind energy systems—it is redefining the entire energy paradigm toward smarter, cleaner, and more resilient futures. As AI continues to evolve, its synergy with wind energy will play a pivotal role in accelerating the transition to global renewable energy goals.

REFERENCES

[1] H. Elmousalami, A.A. Alnaser, and F.K.P. Hui, "Sustainable AI-driven wind energy forecasting: advancing zero-carbon cities and environmental computation", *Artif. Intell. Rev.,* vol. 58, no. 6, p. 191, 2025.
[http://dx.doi.org/10.1007/s10462-025-11191-0]

[2] I. Ul Haq, I.U. Hassan, and H.A. Shah, "Machine Learning Techniques for Result Prediction of One Day International (ODI) Cricket Match", In: *Proceedings of the 2023 IEEE 8th International Conference on Convergence Systems and Innovative Technologies (I2CT)*, 2023.

[3] F. Ekundayo, "Leveraging AI-Driven Decision Intelligence For Complex Systems Engineering", *Int. J. Res. Publ. Rev.,* vol. 5, no. 11, pp. 5489-5499, 2024.
[http://dx.doi.org/10.55248/gengpi.5.1124.3343]

[4] I.U. Haq, A.H. Rather, and G. Kaur, "A Comparative Analysis of Machine Learning Algorithms for the Early Prediction of Cardiovascular Disease", In: *Proceedings of the 2023 2nd International Conference on Edge Computing and Applications (ICECAA)*, 2023, pp. 987-993.
[http://dx.doi.org/10.1109/ICECAA58104.2023.10212214]

[5] S. Kunumalla, D. Rajababu, and A.V.V. Sudhakar, "Evaluation of isolated operation of a wind energy system connected", In: *AI-Driven Solutions and Emerging Power Technologies for Sustainable Future* (ETAEE 2024), A. Yadav, K. M. Silva, and B. R. Bhalja, Eds. Lecture Notes in Electrical Engineering, vol. 1485. Singapore: Springer Nature, 2026, pp. 201–215. [http://dx.doi.org/10.1007/978-981-95-2419-8_17]

[6] A.A. Firoozi, F. Hejazi, and A.A. Firoozi, "Advancing wind energy efficiency: A systematic review of aerodynamic optimization in wind turbine blade design", *Energies,* vol. 17, no. 12, p. 2919, 2024. [http://dx.doi.org/10.3390/en17122919]

[7] F.P.G. GMárquez, and A.P. Gonzalo, "A comprehensive review of artificial intelligence and wind energy", *Arch. Comput. Methods Eng.,* vol. 29, no. 5, pp. 2935-2958, 2022. [http://dx.doi.org/10.1007/s11831-021-09678-4]

[8] I. Ul Haq, "Diagnosis of Neurological Disease Using Bioinspired Algorithms", In: *Bio-Inspired Optimization for Medical Data Mining,* S. Srivastava, A. Anand, A. Kumar, B. Saini, P.S. Rathore, Eds., Wiley-IEEE Press: Hoboken, NJ, USA, 2024, pp. 245-270. [http://dx.doi.org/10.1002/9781394214211.ch12]

[9] K.I. Khan, and A. Nasir, "Application of artificial intelligence in solar and wind energy resources: a strategy to deal with environmental pollution", *Environ. Sci. Pollut. Res. Int.,* vol. 30, no. 24, pp. 64845-64859, 2023. [http://dx.doi.org/10.1007/s11356-023-27038-6] [PMID: 37097570]

[10] S. Nasir, H. Zainab, and H.K. Hussain, "Artificial-intelligence aerodynamics for efficient energy systems: the focus on wind turbines", *BULLET: Jurnal Multidisiplin Ilmu,* vol. 3, no. 5, pp. 648-659, 2024.

[11] I. Ul Haq, G. Kaur, and A.H. Rather, "Fruit Leaf Classification Using Transfer Learning for Automation and Industrial Applications", In: *Deep Learning Techniques for Automation and Industrial Applications,* P.S. Rathore, S. Ahuja, S.R. Burri, A. Khunteta, A. Baliyan, A. Kumar, Eds., Wiley- IEEE Press: Hoboken, NJ, USA, 2024, pp. 115-138. [http://dx.doi.org/10.1002/9781394234271.ch10]

[12] V. Simankov, P. Buchatskiy, A. Kazak, S. Teploukhov, S. Onishchenko, K. Kuzmin, and P. Chetyrbok, "A solar and wind energy evaluation methodology using artificial intelligence technologies", *Energies,* vol. 17, no. 2, p. 416, 2024. [http://dx.doi.org/10.3390/en17020416]

[13] R. Meenal, P.A. Michael, D. Pamela, and E. Rajasekaran, "Weather prediction using random forest machine learning model", *Indones. J. Electr. Eng. Comput. Sci.,* vol. 22, no. 2, pp. 1208-1215, 2021. [http://dx.doi.org/10.11591/ijeecs.v22.i2.pp1208-1215]

[14] I.U. Haq, and A.M.U.D. Khanday, "Exploring the role of explainable AI in Women's health: Challenges and solutions", In: *Explainable Artificial Intelligence in the Healthcare Industry,* A. Kumar, T.A. Kumar, P. Das, C. Sharma, A.K. Dubey, Eds., Wiley-IEEE Press: Hoboken, NJ, USA, 2025, pp. 283-306. [http://dx.doi.org/10.1002/9781394249312.ch13]

[15] N. Goutham, B. Alonzo, A. Dupré, R. Plougonven, R. Doctors, L. Liao, M. Mougeot, A. Fischer, and P. Drobinski, "Using machine-learning methods to improve surface wind speed from the outputs of a numerical weather prediction model", *Boundary-Layer Meteorol.,* vol. 179, no. 1, pp. 133-161, 2021. [http://dx.doi.org/10.1007/s10546-020-00586-x]

[16] H. Liu, X. Mi, and Y. Li, "Smart deep learning based wind speed prediction model using wavelet packet decomposition, convolutional neural network and convolutional long short term memory network", *Energy Convers. Manage.,* vol. 166, pp. 120-131, 2018. [http://dx.doi.org/10.1016/j.enconman.2018.04.021]

[17] I.U. Haq, S.Z. Rufai, A.H. Rather, and H.A. Shah, "Transparency in Disease Diagnosis: Leveraging Interpretable Machine Learning in Healthcare", In: *Explainable Artificial Intelligence in the*

Healthcare Industry, A. Kumar, T.A. Kumar, P. Das, C. Sharma, A.K. Dubey, Eds., Wiley-IEEE Press: Hoboken, NJ, USA, 2025, pp. 112-135. [http://dx.doi.org/10.1002/9781394249312.ch6]

[18] Ł. Mazur, S. Cieślik, and S. Czapp, "Trends in locally balanced energy systems without the use of fossil fuels: a review", *Energies,* vol. 16, no. 12, p. 4551, 2023. [http://dx.doi.org/10.3390/en16124551]

[19] M. Singh, and G. Kaur, "Case Studies on AI-Driven Innovations in Renewable Energy, Waste Management, and Resource Conservation", In: *Maintaining a Sustainable World in the Nexus of Environmental Science and AI,* B. Singh, C. Kaunert, K. Vig, S. Dutta, Eds., IGI Global: Hershey, PA, USA, 2024, pp. 455-484. [http://dx.doi.org/10.4018/979-8-3693-6336-2.ch018]

[20] T. Islam, M. S. Alam, M. G. Muktadir, M. M. Tasnim, J. Islam, and K. N. Sakib, "Siting of Potential Areas for the Sustainable Development of Large-Scale Onshore Wind Farms Using Multi-Criteria Analysis and Geographic Information System: A Case Study on Bangladesh," *Sustainability*, vol. 18, no. 5, p. 2204, Feb. 2026.

[21] K. Mammadov, I. Novruzova, N. Hasanova, and S. Mammadov, "Survey of Machine Learning Applications in Wind Power Implementation: A Comprehensive View", In: *Proceedings of the 2024 IEEE 18th International Conference on Application of Information and Communication Technologies (AICT)*, 2024, pp. 1-8. [http://dx.doi.org/10.1109/AICT61888.2024.10740445]

[22] J. Maldonado-Correa, J. Cuenca-Granda, J. Torres-Cabrera, G. Cerda Mejía, W. D. Bastidas Barragan, R. Guapulema, E. Paccha-Herrera, J. C. Solano, D. Tapia-Peralta, J. Benavides, and C. Laverde-Albarracín, "Digital Twin Technology in Wind Turbine Condition Monitoring, Predictive Maintenance, and RUL Estimation: A Systematic Literature Review," *Energies*, vol. 19, no. 6, p. 1477, Mar. 2026.

[23] N. Sharma, M. Sharma, A. Singhal, R. Vyas, H. Malik, M.A. Hossaini, and A. Afthanorhan, "An efficient approach for recognition of motor imagery EEG signals using the Fourier decomposition method", In: *IEEE Access,* vol. 11, pp. 122782-122791, 2023.

[24] S. Rani, D. Jining, D. Shah, S. Xaba, and P.R. Singh, "Revolutionizing the Creative Process: Exploring the Benefits and Challenges of AI-Driven Art", In: *Intelligent Computing and Optimization*, P. Vasant, *et al*., Eds. Cham, Switzerland: Springer, 2023, pp. 115–128. [http://dx.doi.org/10.1007/978-3-031-50151-7_23]

[25] S. Chinamanagonda, "AI-driven Performance Testing AI tools enhancing the accuracy and efficiency of performance testing. Advances in Computer Sciences, 4(1). Thippanna, G., Jyothi, P. V. S. N., Krishnaiah, P., Thippanna, G., Jyothi, P., & Krishnaiah, P. (2023). An extensive review of the benefits and drawbacks of AI tools", *World Journal of Advanced Research and Reviews,* vol. 20, no. 3, pp. 1668-1673, 2021.

[26] M. Radif, A. H. Al-saeedi, R. R. Nuiaa Alogaili, and M. A. Mohammed, "Enhancing Classification Accuracy through Cluster-Based Ensemble Learning and Adaptive Weighting," *Fusion: Practice & Applications*, vol. 21, no. 1, pp. 131–141, 2026.

[27] E. Gołąb-Andrzejak, "AI-powered digital transformation: Tools, benefits and challenges for marketers–case study of LPP", *Procedia Comput. Sci.,* vol. 219, pp. 397-404, 2023. [http://dx.doi.org/10.1016/j.procs.2023.01.305]

[28] J. Kazak, J. van Hoof, and S. Szewranski, "Challenges in the wind turbines location process in Central Europe – The use of spatial decision support systems", *Renew. Sustain. Energy Rev.,* vol. 76, pp. 425-433, 2017. [http://dx.doi.org/10.1016/j.rser.2017.03.039]

[29] K. Konstas, P.T. Chountalas, E.A. Didaskalou, and D.A. Georgakellos, "A pragmatic framework for data-driven decision-making process in the energy sector: Insights from a wind farm case study",

Energies, vol. 16, no. 17, p. 6272, 2023.
[http://dx.doi.org/10.3390/en16176272]

[30] E.G. Nabati, and K.D. Thoben, "Data driven decision making in planning the maintenance activities of off-shore wind energy", *Procedia CIRP,* vol. 59, pp. 160-165, 2017.
[http://dx.doi.org/10.1016/j.procir.2016.09.026]

[31] P. Pinson, "Wind energy: forecasting challenges for its operational management", *Statist. Sci.,* vol. 28, no. 4, pp. 564-585, 2013.

[32] P. Veers, K. Dykes, E. Lantz, S. Barth, C.L. Bottasso, O. Carlson, A. Clifton, J. Green, P. Green, H. Holttinen, D. Laird, V. Lehtomäki, J.K. Lundquist, J. Manwell, M. Marquis, C. Meneveau, P. Moriarty, X. Munduate, M. Muskulus, J. Naughton, L. Pao, J. Paquette, J. Peinke, A. Robertson, J. Sanz Rodrigo, A.M. Sempreviva, J. C. Smith, A. Tuohy, and R. Wiser, "Grand challenges in the science of wind energy", *Science,* vol. 366, no. 6464, p. eaau2027, 2019.
[http://dx.doi.org/10.1126/science.aau2027] [PMID: 31601706]

[33] M.B. Bouraima, E. Ayyildiz, I. Badi, M. Murat, H.A. Es, and D. Pamucar, "A decision support system for assessing the barriers and policies for wind energy deployment", *Renew. Sustain. Energy Rev.,* vol. 200, p. 114571, 2024.
[http://dx.doi.org/10.1016/j.rser.2024.114571]

[34] P.D. Rigo, G. Rediske, C.B. Rosa, N.G. Gastaldo, L. Michels, A.L. Neuenfeldt Júnior, and J.C.M. Siluk, "Renewable energy problems: Exploring the methods to support the decision-making process", *Sustainability (Basel),* vol. 12, no. 23, p. 10195, 2020.
[http://dx.doi.org/10.3390/su122310195]

[35] S. Dash, S. Chakravarty, N.C. Giri, and R. Khargotra, "Evaluating sustainable wind energy sources with multiple criteria decision-making (MCDM) techniques", *Comput. Electr. Eng.,* vol. 123, p. 110285, 2025.
[http://dx.doi.org/10.1016/j.compeleceng.2025.110285]

[36] A.K. Saha, M. Anjum, Z. Mahmud, and M. Hasanuzzaman, "Wind energy and power plant", In: *Advances in Renewable Energy Conversion Technologies for Power Plant Engineering*, M. Hasanuzzaman, Ed. London, UK: Academic Press, 2026, pp. 475–502.
[http://dx.doi.org/10.1016/B978-0-443-28892-0.00011-6]

[37] R.K. Pandit, D. Astolfi, and I. Durazo Cardenas, "A review of predictive techniques used to support decision making for maintenance operations of wind turbines", *Energies,* vol. 16, no. 4, p. 1654, 2023.
[http://dx.doi.org/10.3390/en16041654]

CHAPTER 3

Machine Learning Models for Resource Optimization in AI-Driven Wind Energy Systems

Inderdeep Kaur[1,*] and **Inzimam Ul Hassan**[2]

[1] *Department of Computer Science & Engineering, University Institute of Engineering, Chandigarh University, Mohali, Punjab, India*

[2] *Department of Computer Science & Engineering, Vivekananda Global University, Jaipur, India*

Abstract: The expanding global need for sustainable energy technologies has made wind energy essential for achieving a renewable energy balance. Wind energy operation faces major difficulties because of its variable natural behavior, which makes forecasting and resource planning complex and affects turbine maintenance operations. The application of machine learning algorithms is transforming wind energy systems through data-driven and intelligent decision systems for resource optimization purposes. This article provides an extensive explanation of major ML models, including Support Vector Machines, Artificial Neural Networks, Reinforcement Learning, Genetic Algorithms, Ensemble Learning, K-Means Clustering, and Long Short-Term Memory (LSTM) networks, which are deployed for forecasting, predictive maintenance, control strategies, and wind farm design applications. This chapter uses Artificial Intelligence (AI) to explain how it modernizes green energy systems by improving operational output while lowering maintenance delays and creating early reactions to variable ecological factors. The chapter presents real-world examples demonstrating how AI effectively maximizes energy yield and reduces operational costs in worldwide wind energy initiatives. Research and deployment of AI receive support through open-source tools that feature datasets and simulation platforms within this domain. Challenges such as data quality, model interpretability, and deployment constraints are critically analyzed, along with future research directions including integration with IoT and smart grids, offshore wind development, and ethical AI practices. By bridging the gap between AI innovation and renewable energy applications, this chapter underscores the transformative potential of machine learning in advancing wind energy toward a smarter, more efficient, and sustainable future.

Keywords: Artificial Intelligence, Machine learning, Predictive maintenance, Resource optimization, Wind power forecasting, Wind energy.

* **Corresponding author Inderdeep Kaur:** Department of Computer Science & Engineering, University Institute of Engineering, Chandigarh University, Mohali, Punjab, India; E-mail: inderdeep.sgtb@gmail.com

Sugandha Singh, Inam Ul Haq, Sachin Ahuja, Akib Mohi Ud Din Khanday & Neha Sharma (Eds.)

INTRODUCTION

Wind power serves as a principal foundation for renewable energy transformation as global efforts to establish sustainable energy systems. Wind power systems face major operational obstacles due to the natural wind, besides dealing with extensive wind farm maintenance requirements [1]. The wind power sector receives transformative changes through Artificial Intelligence because the technology delivers precise prediction tools and enhanced predictive maintenance and operational optimization in real-time, which establishes wind energy as a steady and affordable solution.

IMPORTANCE OF RENEWABLE ENERGY AND THE ROLE OF WIND ENERGY

The global transition to renewable energy is critical to mitigating climate change, reducing greenhouse gas emissions, and ensuring energy security [2]. Among renewables, wind energy has emerged as a leading solution due to its scalability, declining costs, and technological maturity. In 2022, wind power accounted for over 7% of global electricity generation, with projections to reach 20% by 2030 (IRENA, 2023). Wind energy is a key element for achieving the net-zero target because it produces zero direct emissions [3]. The flexible modular system of wind power can be installed in either rural onshore sites or deep-water offshore settings, thereby increasing energy planning adaptability.

Challenges in Optimizing Wind Energy Systems

Despite its promising potential, wind energy faces several obstacles:

- Customer demand suffers due to wind power variability, which makes it harder to operate the grid and determine energy storage solutions.
- Advanced forecasting, along with grid management systems, becomes necessary to integrate wind energy into the electric power grid [4].
- Offshore turbines operating in harsh marine conditions incur higher operational costs, along with the general need for higher maintenance expenses [5].
- Wind turbines installed behind existing facilities experience reduced operational performance because of the turbulent air produced by preceding turbines.
- The repeated failure of turbine components occurs mostly because of their mechanical degradation, such as gearboxes and blades.

FUNDAMENTALS OF WIND ENERGY SYSTEMS

This part introduces foundational wind power generation principles, starting with wind attributes and turbine construction methods and concluding with energy transformation methods. The module introduces the fundamental principles, showing how wind-powered kinetic energy is converted into usable electrical energy through mechanical and electrical devices.

Overview of Wind Energy Generation Process

The conversion of wind energy consists of three main components:

1. **Kinetic Energy Harvesting:** The turbine spins through the force of impinging wind from which kinetic energy is collected.
2. **Mechanical Energy Conversion:** When the spinning rotor activates a shaft together with a gearbox, which increases the speed output to run a generator system.
3. **Electrical Energy Production:** Electrical energy is converted from mechanical energy through the generator and transmitted over power lines into the grid.

Key Components of Wind Energy Systems

- **Turbines:** Horizontal-axis wind turbine (HAWT) and vertical-axis wind turbine (VAWT) designs, with HAWTs dominating utility-scale projects.
- **Sensors:** Monitor wind speed, direction, temperature, vibration, and torque (*e.g.*, anemometers, lidar, accelerometers).
- **Controllers:** Adjust blade pitch and yaw angle to maximize energy capture and minimize mechanical stress.
- **Grid Interface:** Inverters and transformers synchronize turbine output with grid frequency and voltage.

Types of Wind Farms

a. Onshore: Installed on land; lower installation costs but face challenges like land-use conflicts and lower wind consistency [6].
b. Offshore: These facilities, built in ocean locations, can capture stable winds at the cost of greater financial outlay and deteriorate from corrosion [7].
   - Fixed-bottom: Suitable for shallow waters (≤60m depth).
   - Floating systems can operate in deep waters by using mooring lines as an emerging technology.

Resource Optimization Challenges in Wind Energy

- **Spatial Optimization:** Balancing turbine density to minimize wake losses while maximizing energy yield [8].
- **Temporal Variability:** Aligning energy production with demand cycles (*e.g.*, peak evening hours).

- **Material Constraints:** Ensuring component durability in extreme weather (*e.g.*, offshore saltwater corrosion).
- **Economic Trade-offs:** Weighing upfront costs (*e.g.*, floating turbines) against long-term energy gains.

EMERGENCE OF AI AND MACHINE LEARNING

The combination of ML and AI systems brings revolutionary changes to energy systems through their data-centric decision processes. ML models analyze enormous data sets comprising meteorological records, turbine sensor data, and grid data to achieve operational excellence in wind power plants [9]. Applications include:

Enhancing Forecasting Precision

The irregular flow of wind energy creates significant challenges for both power grid connection and power storage system development. Historical data-based and simple meteorological model-dependent forecasting methods demonstrate a poor capacity to detect complicated atmospheric conditions [10]. Machine learning algorithms in Artificial Intelligence analyze multiple datasets containing wind speed, direction, pressure, and temperature information to deliver precise predictions. The application of LSTM networks allows processing of time-series data to forecast upcoming short-term changes, although ensemble models unite different algorithm predictions to cut off modeling mistakes [11]. Using neural networks, DeepMind optimized a 700 MW wind farm and increased energy delivery value by 20% by improving grid scheduling. By improving predictions, the power grid becomes more stable and uses less fossil fuel backups when wind generation decreases, which helps speed up the process of decarbonization [12, 30].

Predictive Maintenance: From Reactive to Proactive

The practice of using fixed schedule-based maintenance and reactive repairs generates high expenses through system downtime and needs premature replacement of equipment. AI-based predictive maintenance utilizes a turbine SCADA system and vibration sensor, and thermal camera inputs to detect component wear events as well as upcoming failures. Real-time operational data analysis through RF and anomaly detection algorithms helps ML models to estimate component lifespan values. Siemens Gamesa applies digital twins consisting of virtual turbine duplicates to run simulation tests that discover system faults before actual weeks elapse [13]. The maintenance costs decrease by 30% while the turbine lifetime extends through this approach, which guarantees consistent energy production.

Optimal Control Through Adaptive Learning

Artificial Intelligence deploys dynamic operations for turbines to reach top energy capture efficiency, together with reduced mechanical stress. The reinforcement learning component of RL algorithms uses trial and error methods to find optimal strategies that control the blade pitch angles together with the yaw angles through immediate wind pattern recognition. AI-empowered devices powered by Vestas's Smart Yaw Control system enable machines to steer turbines toward special directions, thus allowing downstream units to escape wake effects and improving total efficiency by 5–10%. Through its accurate control systems, the system raises annual energy production while reducing maintenance expenses of the gearbox components over extended periods [14].

Data Synthesis and Technological Integration

The functionality of AI depends on its capacity to analyze enormous datasets, which consist of meteorological station and turbine sensor data, and satellite imagery data. The collection of data in real time originates from IoT devices at the edge that transfer the information to cloud processing platforms, where extensive analytics takes place. A model's accuracy depends on feature engineering methods because these methods remove turbulence intensity along with gearbox temperature from noisy information sources [15]. The combination of AI solutions with 5G technology and digital twin systems enables developers to achieve better prediction abilities through a cyclic improvement process of their continuous modeling activities.

Importance of Data-Driven Decision-Making in Wind Energy

Lawful data collections stem from sensors that gauge meteorological conditions, besides measuring turbines' operation and grid system connections. The mass data becomes interpretable through Artificial Intelligence systems, allowing developers to support decision-making processes [16]. Strategies based on data optimization enable critical wind turbine decision support for site identification while also balancing network systems and managing capacity [17]. AI serves as the fundamental driving force to achieve worldwide clean energy transition by improving wind energy system sustainability through efficiency increases, as well as economic stability.

AI applications have introduced a revolutionary period in wind industry management. Customers within the sector overcome their principal challenges by implementing AI methods, which improve predictive maintenance capabilities and enhance turbine performance predictions and forecasting precision [18]. Better worldwide progress towards net-zero emissions will result from advances in algorithms, which will lead to wind energy's competitiveness with traditional sources. Wind technology demonstrates how machine learning systems attempt to transform both clean renewable energy generation and make wind power into our essential base foundation.

MACHINE LEARNING MODELS FOR RESOURCE OPTIMIZATION IN WIND ENERGY SYSTEMS

Maximum operational efficiency together with sustainability in wind energy systems depends on resource optimization measures, which enable cost reductions and operational benefits. The power revolution depends on ML models to make data-driven decisions, which enhance energy output and reduce operational budgets as well as environmental impact. The analysis of median performance uses combined data points from climate factors and turbine operations, and power grid consumption thresholds in combination with maintenance needs, to bring innovative approaches to wind energy resource distribution.

Key Applications of ML in Resource Optimization

a. **Wind Farm Layout Design:** The site placement of wind turbines determines energy generation rates because the wind flow disturbance between neighboring turbines disrupts downstream power production. Traditional layout construction bases its designs on simplified fluid dynamics modeling systems that dismiss both effects of turbulence, together with variations in terrain shape [18]. The wind farm optimization process driven by ML models, including both CNNs and GAs, allows turbine position optimization *via* fluid flow simulation across multiple operational scenarios. Through its Mahimn project, Google applies reinforcement learning techniques to reduce wake effects, which boosts complete farm performance by 15 percent. Digital Wind Farm platform from GE enables performance optimization through predictive analytics, resulting in a 20% increase in annual energy production (AEP).

b. **Dynamic Load Distribution:** The operation of wind turbines exposes them to variable mechanical tensions produced by shifting wind speed patterns and directions. Random forests and gradient-boosted trees evaluate sensor data, including vibration, torque, and temperature, to forecast component fatigue so that dynamic load redistribution becomes possible. Through real-time machine learning algorithms, the Loads and Control Optimization system from Vestas modifies the blade pitch along with rotor speed to simultaneously optimize power output and component durability. The incorporation of this system leads to life extension of turbines while cutting maintenance expenses by 12–18 percent [19].

c. **Grid Integration and Energy Storage:** The connection of intermittent wind power to the power grid demands the exact timing of supply to match consumer requirements. The forecasts generated by ML models enable the optimization of battery storage operations combined with grid stability tasks. DeepMind uses neural networks to predict upcoming wind power availability for 36 hours, which helps utilities operate battery storage systems efficiently [20]. The unsupervised learning method known as clustering helps determine the most favored instances for storing excess energy and selling it commercially to achieve maximum profit.

d. **Material and Logistics Optimization:** The utilization of resources within aviation becomes smoother through the application of ML in business operations. The predictions of supervised learning models determine proper timings for replacing parts, including gearboxes and blades. The utilization of natural language processing technology by SparkCognition allows businesses to analyze maintenance logs, which leads to the prioritization of high-risk turbines and results in a 25% reduction of downtime [21]. Using reinforcement learning enables optimal logistics route planning for technicians so their operations become more fuel-efficient and environmentally friendly.

Dominant Machine Learning Techniques

- **Supervised Learning:** The performance of SVMs and neural networks in regression and classification scenarios enables them to forecast turbine malfunction occurrences and estimate energy production outcomes from past measurements.
- **Unsupervised Learning:** Clustering algorithms (*e.g.*, k-means) group turbines with similar performance profiles, enabling targeted maintenance and resource allocation.
- **Reinforcement Learning:** RL agents learn optimal control policies through trial and error, such as adjusting yaw angles to maximize energy capture while minimizing wear.

Machine Learning Models for Resource Optimization

Below are some machine learning models used for the optimization of resources:

Support Vector Machines (SVM)

The supervised learning system, Support Vector Machines (SVM), functions as a method for classification along with regression tasks [22]. SVMs have gained high acceptance in wind energy systems for wind speed and power output prediction through their data learning process, which processes historical meteorological datasets including wind direction data alongside air pressure data, temperature readings, and previous speed measurements. The predictive models become more accurate through this method, which helps operators schedule future energy dispatches. The use of SVM stands out because its kernel functions enable successful operation with high-dimensional data as well as non-linear relationships. As illustrated in Fig. (**1**), the Support Vector Machine (SVM) algorithm identifies the optimal hyperplane that maximizes the margin between classes.

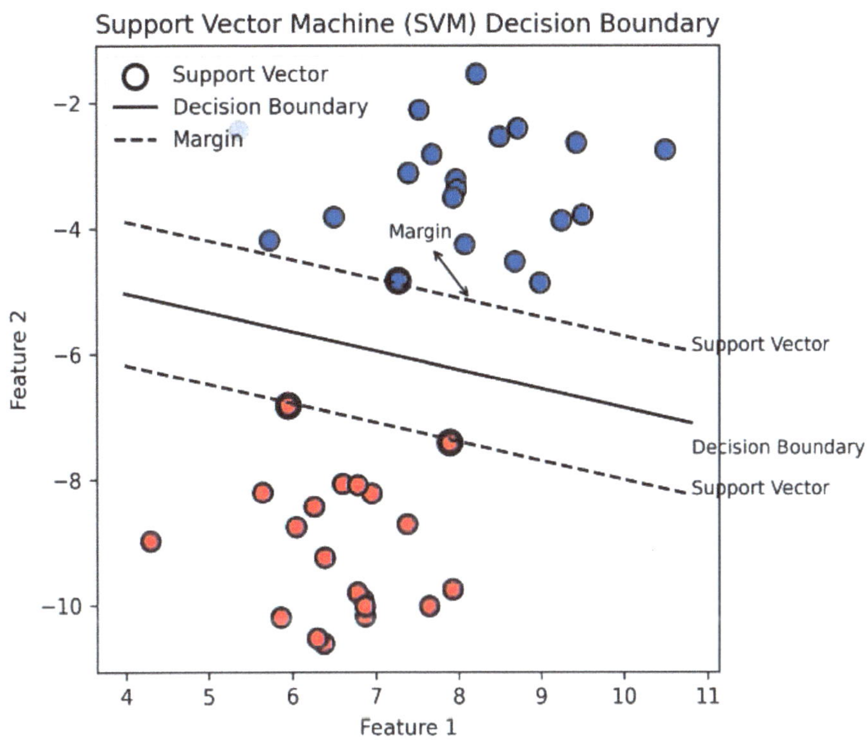

Fig. (1). Support vector [22].

VMS-based accurate wind power forecasting contributes to minimizing administrative challenges in the scheduling of turbines and their integration into power grid operations. Accurate predictions of wind energy generation help minimize reliance on non-renewable backup resources and improve the management of wind resource utilization. SVM contributes to diminished operational expenses and decreased energy losses, as well as enhanced production of sustainable energy infrastructure systems. The combination of SVM technology is essential for resource optimization since it supports well-informed decisions that enhance the operational effectiveness of wind energy systems.

Artificial Neural Networks (ANN) / Deep Learning

The artificial neural networks, together with deep learning methods, imitate brain neural networks for analyzing complicated data patterns within extensive datasets [23]. ANN technology demonstrates significant effectiveness for analyzing various complex connections between wind energy variables like wind speed direction, turbine data indicators and meteorological conditions. The detection of obscure patterns and trends by Convolutional Neural Networks (CNNs) and Recurrent Neural Networks (RNNs) represents one of the most important uses of ANN in wind energy. The processing of turbine sensor information through ANN models enables the detection of initial evidence indicating mechanical decline, together with performance deterioration. The acquired insights enable proper planning of maintenance sessions, which helps prevent equipment breakdowns and decrease operational downtime. Through this method, maintenance functions in a structured way, which leads to frequent power output stability, longer equipment component operational life and reduced repair expenditures. The architecture in Fig. (**2**) is particularly well-suited for processing large-scale meteorological datasets in wind energy forecasting.

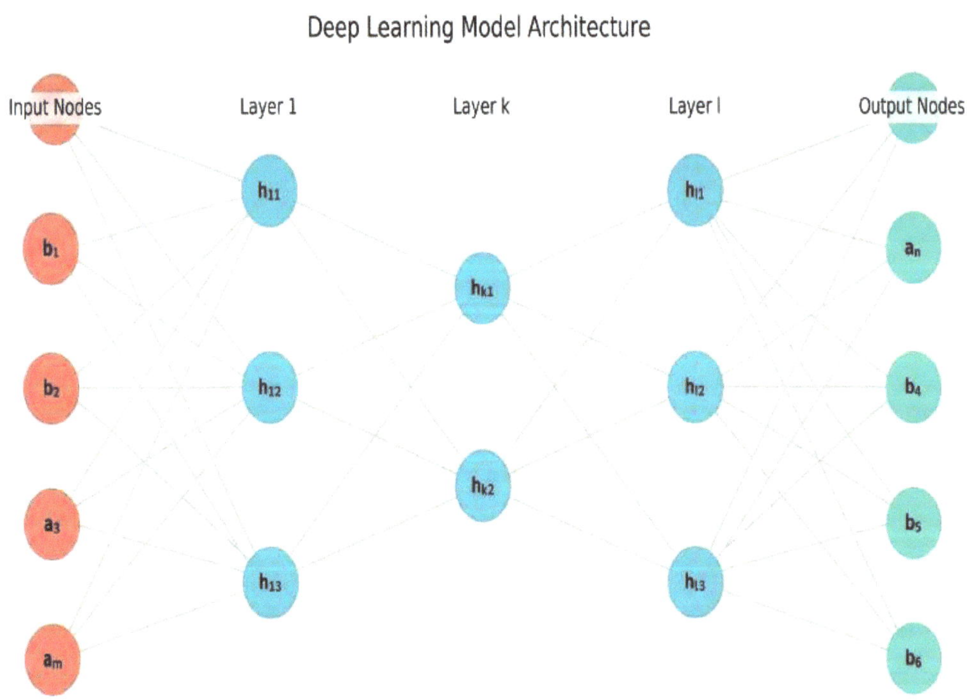

Fig. (2). Deep learning architecture [23].

Wind pattern analysis strongly relies on ANNs for its operations. Predictive models generate exact wind pattern results throughout time, which enables renewable energy installations to position turbine towers and boost their energy performance. Deep learning models learn accuracy through real-time data processing, and at the same time, they stay adaptable toward developing environmental conditions throughout their self-improvement process. Higher operational performance, combined with superior energy generation, happens along with reduced operational expenses because of this achievement.

Reinforcement Learning (RL)

Agents operate within reinforcement learning as a machine learning method to discover environmental responses after direct interaction by obtaining either feedback through reward or punishment [24]. The wind energy operations have witnessed an increased adoption of RL for live turbine management and dynamic energy distribution. RL surpasses standard control systems since it enables turbines to achieve superior operational strategies through dynamic testing of actions and assessment of outcomes for power output results. Wind patterns direct operating agents to optimize turbine blade positioning as well as attack position, which results in increased energy capture efficiency. Through its operations, the system enhances power output levels and prolongs mechanical system durability, which affects turbine lifespan. As depicted in Fig. (**3**), reinforcement learning involves an agent interacting with an environment to learn optimal actions through rewards and penalties.

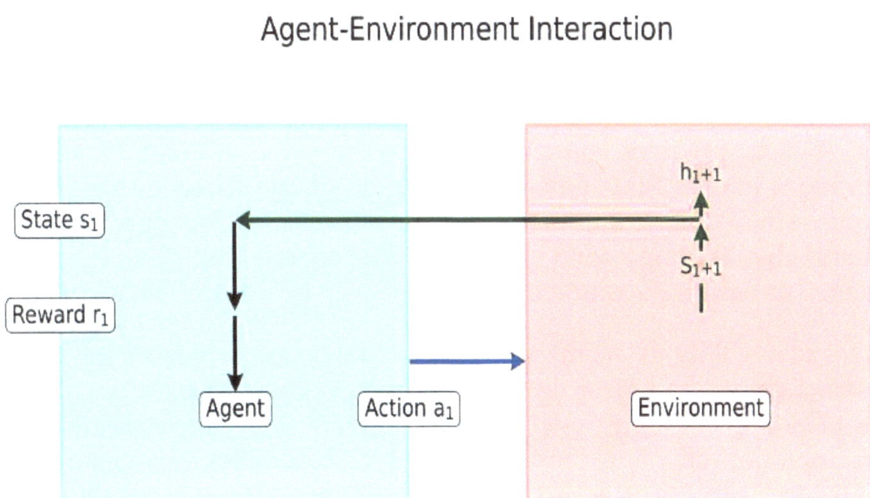

Fig. (3). Reinforcement learning [24].

An RL algorithm manages the distribution of resources across wind turbines throughout wind farms through its evaluation of environmental factors such as wind speed variations, while considering turbine health parameters as well as power specifications. The machine learning model enables system policy creation, which delivers high energy efficiency along with operational endurance without jeopardizing equipment maintenance requirements [25]. Reinforcement Learning provides excellent solutions for wind energy, mainly because it adapts well to changes in environmental conditions.

Gravity Energy has identified a promising wind energy application that integrates RB with electric grids in distributing power while adapting storage processes to shifting wind production levels. The system improves power reliability by maintaining stability levels and stops energy decreases to deliver steady power. Wind energy producers benefit from Reinforcement Learning because this learning method develops efficient optimization tools that enable data-driven autonomous control systems. Real-time learning abilities make this tool a valuable instrument that produces maximum energy at reduced costs with enhanced resilience to green energy infrastructure.

Genetic Algorithms (GA)

GA represents an optimization methodology that uses biological natural selection principles to develop better solutions by executing selection and crossover, followed by mutation [28]. GA has established itself as an important optimization method for wind energy systems to improve wind turbine placement and farm array design within wind farms. To determine suitable turbine placements, multiple factors must be considered, which consist of terrain characteristics coordinated with wind patterns and impacts on downstream turbine speeds from upstream wind turbines, and the need to follow environmental restrictions. The process of GAs involves numerous fitness function evaluations of potential configurations until it finds placements with the best efficiency for minimizing interference energy while improving system output [26]. GAs operate as adaptive instruments by uniting wind resource assessment tools with geographic information to design suitable solutions for their evaluation sites.

GA offers adaptability as the most significant advantage within its set of features. GA systems grant wind farm designers access to this algorithm for their initial process planning phase, as well as for continuous redesigns, which adapt to weather condition changes, land management requirements, and infrastructure needs. Due to its ability to adjust to new requirements, GA establishes itself as a powerful resource in wind energy development processes. Wind energy resource optimization takes place by using Genetic Algorithms because these algorithms develop intelligent data-driven solutions that optimize turbine placement and system constructions. Genetic algorithms solve complex multi-variable problems to enhance energy output while cutting costs at the same time and providing continuous sustainability to wind farm operations [27].

Ensemble Learning (Random Forest, Gradient Boosting)

Ensemble learning techniques, Random Forest and Gradient Boosting, assemble various base learners to produce more precise and generalized results than any independent model would produce [28]. These methods enable optimal performance in wind energy forecasting, particularly through their application for selecting essential features and predicting power output. Random Forest constructs multiple decision trees during training, then produces predictions by obtaining the mean value from its component trees. Such a modeling technique prevents overfitting problems while improving the model's resistance to errors. This method delivers optimal results by identifying the most important meteorological and operational aspects affecting wind energy generation, which include wind speed, air pressure, humidity, and turbine operational status. The input space becomes easier to manage with Random Forest because its feature importance measurements create a faster training process and better prediction performance. As illustrated in Fig. (**4**), ensemble learning combines the predictions of multiple base models to improve overall performance and reduce overfitting.

Gradient Boosting constructs models one after another because each added model serves to refine the mistakes of prior models. The method succeeds at detecting difficult, complex non-linear linkages between variables in wind energy datasets. The forecasting of wind power generation at high precision levels is essential for power grid supply-demand balance, and the operators use accurate outputs to allocate resources more effectively by minimizing the need for fossil-fuel backup and limiting curtailed energy. These learning techniques demonstrate both scalability to different data environments and adaptability, which makes them appropriate for instant deployment across various wind farm settings [29]. Resource optimization becomes possible through wind energy forecasting improvement, which results from Random Forest and Gradient Boosting implementation, thus improving energy distribution and grid management decision-making effectiveness.

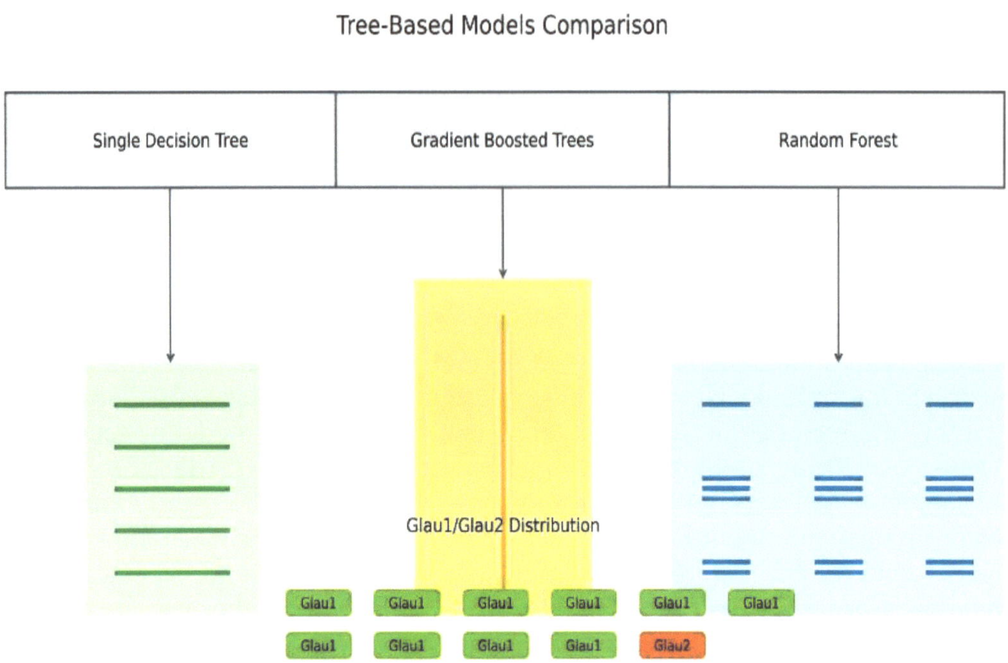

Fig. (4). Ensemble learning [28].

K-Means Clustering

K-Means Clustering serves as an unsupervised learning algorithm that provides highly effective results when grouping data points according to matching features into clusters [30]. The K-Means technique finds extensive application in wind energy and helps identify wind farm sites, process meteorological data, and detect local wind patterns. The main mission involves helping with wind farm planning by identifying how wind data is distributed across space and time. Basing decisions about wind farm development on suitable areas requires evaluation of multiple environmental conditions, including yearly wind velocity levels and direction changes, together with height above sea level and nearby power system infrastructure locations. Through K-Means clustering of sites based on their characteristics, the decision becomes easier for choosing suitable locations. This method helps businesses avoid high-risk locations when they focus their resources on sites that will produce optimal results. The K-Means algorithm leads to valuable results for predicting weather, along with operational planning activities. Wind pattern detection across various seasons and times enables operators to receive vital information that assists them in developing wind-related operational plans. As shown in Fig. (**5**), K-Means clustering partitions the dataset into distinct groups based on similarity, using centroids to define each cluster. The clustering algorithm detects atypical patterns, which enable organizations to enhance predictive maintenance by detecting abnormal operational behaviors.

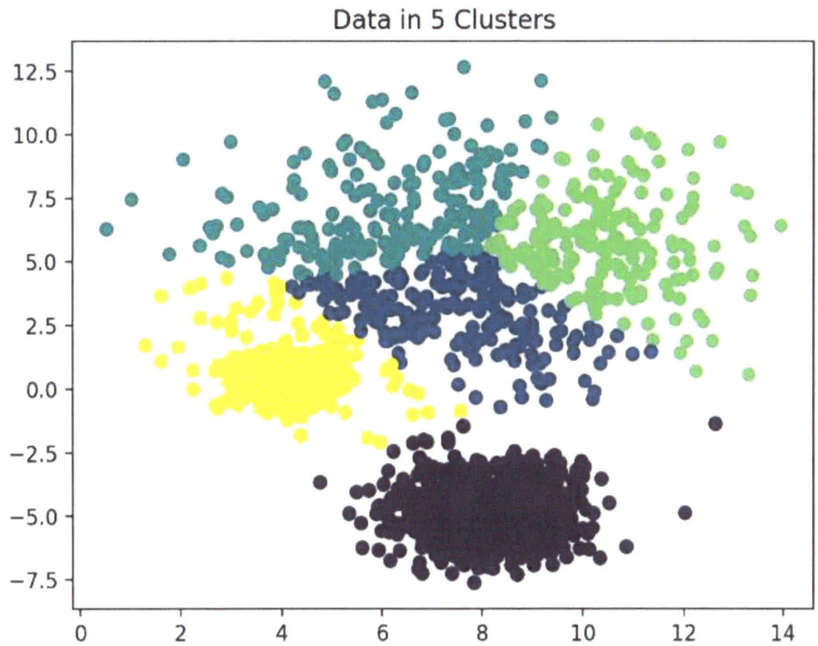

Fig. (5). K means clustering [30].

The resource optimization capabilities of K-Means include facilitating efficient data analyses that improve turbine placement decisions, together with supporting dynamic operational modifications. The algorithm achieves significant value through its capability to transform large data collections into understandable clusters, which enhances the whole wind energy data processing. K-Means Clustering acts as a critical element of AI-powered green energy solutions because it helps planning processes and boosts forecasting models, and boosts the efficiency and effectiveness of wind energy projects.

Long Short-Term Memory (LSTM) Networks

Recurrent Neural Networks (RNN) have Long Short-Term Memory (LSTM) as their subtype because this variant works on sequential data, along with maintaining lengthy relationship maps [31]. The time-series forecasting process for both wind speed parameters and power output measurements in wind energy systems makes extensive use of LSTM networks. Extensive temporal patterns produce effective execution results from LSTM networks, which earns them value in difficult climate sequence processing. Accurate, precise forecasting of power demands in wind energy systems is critical for both maintaining balanced electric supply system operations and storage system activities. High precision comes from LSTM networks because their learning process relies on previous wind data. Wind farm management benefits from exact estimates about energy storage duration when deciding how much power to transfer to the grid and adjusting turbine operations before weather transitions. Fig. (**6**) presents the internal architecture of an LSTM unit, highlighting the gate mechanisms that enable the model to retain or discard information over time.

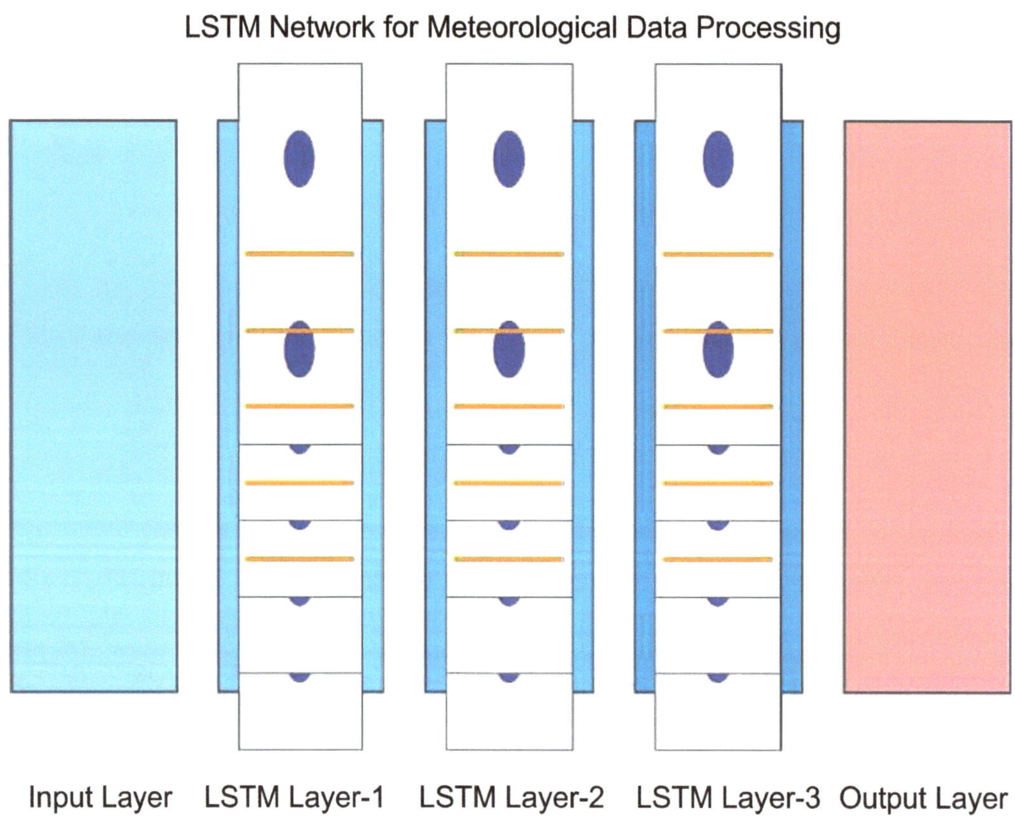

Fig. (6). LSTM architecture [31].

The application of LSTMs allows power operators to improve power grid connection capabilities by providing advanced predictions for wind variability over brief periods. Higher levels of wind power integration require stability improvements because electrical grid control becomes challenging with these conditions. Short and medium-term forecasting outcomes from LSTMs lead to better power grid stability since they reduce the need for backup power systems. The best results from LSTM models emerge through their combination with attention processes and hybrid models constructed from CNNs to assess spatial data patterns. The models achieve improved functionality and gain wider operational capabilities when used for wind energy analytics. Time-series data training of LSTMs results in continuous model enhancement because the algorithms learn new wind patterns to achieve better forecasting precision. The application of LSTM networks delivers essential benefits to predict time-series data in wind energy systems. The prediction functions enable more efficient energy storage operations and improve both turbine scheduling and grid integration strategies. The AI-powered system incorporates LSTM to reach its highest level of efficiency, reliability, and sustainability when handling wind energy production.

Hybrid and Advanced AI Techniques in Wind Energy Systems

Sophisticated Artificial Intelligence techniques are required for wind energy systems because of their technical complexity to achieve robustness, together with adaptive performance and operational efficiency. Wind farm operation undergoes transformation through the application of hybrid and advanced techniques that include multi-model integration, together with transfer learning, federated learning, and edge AI.

a. **Combining Multiple Models for Enhanced Performance:** Hybrid AI frameworks combine multiple different algorithms that support each other in order to address individual model constraints. Wind forecasting accuracy increases through ensemble learning, which uses neural networks as well as gradient-boosted trees together with physics-based models [32]. DeepMind at Google incorporated LSTM networks with NWP data to create a hybrid forecast model, which delivered 20% less forecast errors while achieving higher grid scheduling efficiency. Fluid dynamics equations become an integral part of neural networks through a physics-informed machine learning framework, which allows precise modeling of wake effects to determine turbine locations. Hybrid models adopted by Vestas allow them to maximize annual energy production by 10% while preserving component durability.
b. **Transfer Learning and Federated Learning:** The adaptive capability of transfer learning lets knowledge from large datasets located in onshore farms automatically learn new settings found in offshore areas with little model

training requirements. Siemens Gamesa utilized the method to multiply predictive maintenance model applications between European onshore turbines and Asian offshore installations, which reduced deployment time by 40%.

c. **With federated learning:** distributed model training occurs between independent turbines that maintain their sensitive data. Global wind operators could accomplish distributed data insights aggregation across their different geographic regions through federated learning while maintaining the privacy of sensitive data. The methodology minimizes communication expenditures while keeping data in line with area-specific governance requirements that matter during global turbine operations.
d. **TinyML and Edge AI for Real-Time Optimization:** The incorporation of TinyML on turbines allows direct deployment of lightweight AI models for making real-time edge-based decisions. Local AI controllers use the installation of neural networks on the device to operate blade pitch and yaw angles for quick reactions to sudden wind gusts. The bearing failure prediction system from GE operates through local TensorFlow Lite models, which run on microcontrollers to perform identification without relying on cloud systems or experiencing latency delays. This enables offshore farms to gain substantial benefits from edge device processing of sensor data (vibration and temperature) in real time, which results in automatic maintenance alerts without large-scale data transmission. The anomaly detection capabilities of TinyML enable SparkCognition to lower operational downtime in remote facilities by 25% as a startup company.
e. **Synergy and Impact:** The methods combine to resolve wind energy's fundamental problems, which include intermittent power generation and high operational expenses, along with isolated data systems [33]. The combination of hybrid models together with transfer learning techniques, federated learning methods, and edge AI deployment improves both reliability and speed of forecasts. Advanced AI methods find multifunctional importance in wind energy development toward efficiency and sustainability goals that will transform wind power into a strategic smart grid component. Table **1** summarizes key ML techniques, their respective application areas, and the corresponding optimization goals targeted in each case.

Table 1. ML techniques and their applications with optimization goal.

ML Technique	Application	Optimization Goal
Support Vector Machines (SVM)	Forecasting wind speed/power output.	Helps schedule turbine operation for efficient use of resources.
Artificial Neural Networks (ANN) / Deep Learning	Predictive maintenance, load forecasting, and wind pattern recognition.	Minimizes downtime and maximizes energy output by learning from historical data.

(Table 1) cont.....

ML Technique	Application	Optimization Goal
Reinforcement Learning (RL)	Turbine control, dynamic resource allocation.	Learns optimal policies for turbine adjustments based on real-time data.
Genetic Algorithms (GA)	Optimal placement of turbines, system configuration.	Finds the best configurations for maximizing energy capture.
Ensemble Learning (Random Forest / Gradient Boosting)	Feature selection and predictive modeling of power output.	Accurately predicts wind energy generation to balance supply and demand.
K-Means Clustering	Wind farm site classification, grouping weather patterns.	Helps in planning and designing wind farms by clustering similar data.
LSTM Networks	Time-series forecasting of wind speed and power.	Helps with energy storage and grid integration strategies.

CASE STUDIES AND REAL-WORLD APPLICATIONS OF AI IN WIND ENERGY

Case studies and real-world applications of AI in wind energy highlight how organizations use AI and machine learning to improve wind forecasting, optimize turbine performance, and reduce maintenance costs. These examples demonstrate the practical impact of AI in enhancing the efficiency and reliability of wind energy systems.

1. **Case Study: Google's DeepMind and Wind Energy Forecasting** [34]
   DeepMind, as a Google subsidiary, worked together with Alphabet to enhance energy generation performance in U.S. Midwest wind farm operations. The challenge arose because the unpredictable nature of wind makes it difficult to integrate with the grid, while lowering the economic worth of wind energy caused by variable electricity output. DeepMind employed deep neural networks trained on historical weather data (*e.g.*, wind speed, direction, pressure) and turbine performance records. The models predicted wind power output 36 hours in advance with granular accuracy. A key innovation was the use of time-series forecasting techniques, including Long Short-Term Memory (LSTM) networks, to capture temporal patterns in wind behavior. The predictions then synchronized with grid scheduling throughout the system. The project enhanced the wind energy economic value by 20% due to better time-to-time power sales and decreased need for backup fossil fuels during wind power fluctuations. Through better scheduling procedures, the AI system proved its ability to change irregular energy sources into dependable utility assets while improving power grid stability.

2. **Case Study: Siemens Gamesa Renewable Energy and Predictive Maintenance** [35]
   Siemens Gamesa, which positions itself as a leading wind turbine manufacturer, needed to reduce high operational expenses that occurred due to unexpected turbine breakdowns, coupled with their traditional reactive maintenance methods. The company adopted AI-powered predictive maintenance supported by neural networks and digital twin technology. Many analytical sensors from inside turbines maintained continuous observation of vibrations together with temperature measurements and fluid lubrication conditions. They aim to detect deviations in data while providing advanced warnings about component breakdowns (such as bearing deterioration and blade structural issues) that occurred within weeks ahead. The virtual copies of physical turbines known as digital twins allowed researchers to test various operational conditions virtually, thus improving their predictions [36]. The implemented system reduced unplanned downtime by 30% simultaneously lowering maintenance costs by 15–20% respectively. The transition to proactive maintenance methods at Siemens Gamesa allowed the company to better schedule technicians while achieving longer turbine service life through the effective utilization of Artificial Intelligence.
3. **Case Study 3: Vestas Wind Systems and Ensemble Learning for Power Forecasting** [37]
   Vestas worked to enhance power output prediction methods for optimizing turbine installation and power grid placement, mostly in areas showing challenging terrain and climate variations. The company combined predictive models from random forests and neural networks together with gradient-boosting machines (GBM). The integrated predictive system brings meteorological data such as satellite weather patterns, lidar measurements, with turbine performance measurements. Enhancement through ensemble modeling reduced specific bias in forecasts, which improved accuracy up to 10-15% as compared to single model prediction systems. Dynamic supply-demand balancing by grid operators relying on predictions from the system resulted in an 8% decrease in curtailment losses. Vestas proves through their solution that combined Machine Learning models enhance both system energy production and power grid operational stability [38].

TOOLS, DATASETS, AND PLATFORMS

The successful application of machine learning to optimize wind energy operations depends completely on strong analytical tools and dependable datasets combined with advanced simulation capabilities. The underlying base forms the basis to create, test, and deploy intelligent systems for green energy generation.

a. **Open-source Tools and Frameworks:** Various free tools exist for the efficient execution of machine learning algorithms. Platform developers utilize TensorFlow and PyTorch as deep learning libraries because these tools provide adaptability along with extensive capabilities for creating neural networks that achieve predictive maintenance objectives, forecasting tasks and turbine control applications. Machine learning practitioners use Scikit-learn because it provides a detailed set of classical models, including Support Vector Machines (SVM) and Random Forests, together with clustering algorithm functionality [39]. The data preprocessing chain and analysis functions depend on three essential Python libraries, which include Pandas, NumPy, and Matplotlib. The tools enable researchers and engineers to create proprietary models and evaluate them through various situations for improved performance enhancement.
b. **Public Wind Datasets:** ML models require high-quality datasets to conduct their training and evaluation procedures. The National Renewable Energy Laboratory (NREL) maintains free access to wind speed direction and power output measurement data through its spatially diverse datasets. ECMWF operates ERA5 as a dataset for hourly estimates of global atmospheric variables, which also contains wind-related measurements. In addition to these sources, researchers can utilize dataset collections from both the Open Power System Data (OPSD) and Wind Integration National Dataset (WIND) toolkit. The gathered datasets help model creation for general applications while creating conditions to evaluate different solution approaches.
c. **Simulation Environments for Wind Farms:** Simulation software acts as an essential tool for predicting wind farm performance conducting optimization tests, and validating AI solutions before commercial usage. The assessment tools OpenWind, along with FAST by NREL and WindPRO, give researchers complete control over wind flow evaluation and turbine positioning studies and energy output predictions in virtual environmental conditions. Once integrated with AI models, these environments can evaluate AI models to determine their performance capabilities in real-life conditions. The pair of software applications, including MATLAB/Simulink and HOMER Energy, allows modeling control systems and optimizing microgrids.

The wind energy sector benefits from ML-powered solutions that developers produce with open-source tools, publicly available datasets and high-fidelity simulation environments.

CHALLENGES

The main issue with using machine learning for wind energy systems stems from unreliable data availability. The implementation of wind energy depends critically on meteorological data at both historical phases and real-time periods, as well as turbine operational measurements and power grid performance records. The availability of data for wind energy studies faces operating impediments due to privacy restrictions, data ownership limitations, and incomplete and inconsistent dataset situations. Model performance and generalizability remain dependent on the reliability of training data because sensor malfunction and missing values, together with data noise, affect training data quality. The various data collection practices across regions create obstacles for researchers when developing models that can be used globally because they prevent research scalability and deployment. The process of wind energy optimization faces significant barriers because several AI models used in this field lack easy human interpretation and generate distrust among users.

The complex deep neural networks together with ensemble techniques and reinforcement learning operate through black box mechanisms, which makes it hard to determine the derivation methods for decisions. The absence of information about decision-making processes becomes an adoption barrier, especially for vital applications running real-time turbine control and grid stability management alongside safety systems. Mission-critical utility operations depend on engineers and regulators, along with utility companies, to utilize untransparent algorithms because they need to see the inner workings of these systems. There emerges a substantial need in this field for interpretable and explainable AI (XAI) frameworks due to these trends [40]. The high computational requirements, together with deployment problems, create substantial difficulties within remote and offshore wind installations. Leading AI algorithms need considerable computational power through GPUs or cloud systems, and their implementation might exceed the capacity of real-time applications. Software-hardware solutions that deliver efficient performance for real-time processing must be developed when deploying models at strict on-site computing locations. AI application at the device and embedded system level requires lightweight models built using compression approaches, alongside quantization techniques and with TinyML methods.

FUTURE DIRECTIONS AND RESEARCH OPPORTUNITIES

AI implementation alongside IoT technology and Smart Grid systems provides a beneficial approach to enhancing wind energy solutions. Wind turbine systems with IoT technology, along with sophisticated sensors, gather current performance and weather data as well as construction health information for transmission to control centers. The processing phase of AI algorithms enables them to extract valuable information from collected data in order to make predictions about maintenance requirements, energy output, and grid connectivity. AI assumes a core position in smart grids to control energy storage operations and balance power requirements for strengthening the complete distribution system. The unified ecosystem helps improve operational performance and response rate capabilities across all components of renewable energy operations.

Present-day AI investigations explore offshore wind projects and floating wind park systems because they enable the future growth of wind power and reduce nesting challenges. Managing operations in offshore areas demands tackling two critical problems that include harsh weather elements alongside challenging logistics and expensive maintenance provisions. AI technology provides two operational abilities: automated monitoring, along with fault detection and prediction analytics systems, which address these issues. Programmed intelligence systems allow wind platforms to combine ocean condition monitoring with platform stability assessments using artificial learning models that optimize performance and maintain structural stability. The developed innovations serve as essential infrastructure for expanding offshore wind energy capacity across the globe.

The creation of machine learning systems for wind energy depends primarily on ethical AI principles as well as sustainability standards. Real-world infrastructure depends on the adaptive models, thus requiring full assessment of their sustainable, transparent, and equitable operation in the field. Research professionals should develop ethical standards to protect user confidentiality through unbiased decision-making while establishing rules for transparency as well as accountability procedures. The goal of Green AI is to enable environmentally sustainable operations of AI systems by developing more efficient computational design while using models that require reduced energy expenditures. Science researchers must create distributed training techniques to reduce information transfer throughout the development of power-efficient AI systems for wind-based power generation.

CONCLUSION

Machine learning algorithms function as critical solutions for wind energy resource optimization, which creates a data-based smart framework to oversee energy resource handling. Algorithms established for wind energy prediction reach greater accuracy by processing current measurements together with past data records from which the system derives commands for turbine management operations and implements wind farm design strategies while creating predictive maintenance schedules. The combination of Support Vector Machines, Artificial Neural Networks, Reinforcement Learning, and Genetic Algorithms-based solutions leads to better reliability and efficiency for wind energy production by identifying and implementing complex decision patterns. Lower operational expenses combined with diminished power loss enable easy renewable energy integration into electric power networks, which represents the key role of wind energy resource optimization. Machine learning systems operate such that they adjust their functions while detecting predictive equipment failure indicators, thus enabling resource-based decisions by monitoring real-time operational needs. The combination of ensemble learning with deep learning frameworks expands solution strength and usability because they process large multidimensional datasets to unite various insights. The long-term advantages of applying machine learning to optimize resources challenge technical obstacles based on obtaining data and making it understandable, along with coping with computational resource needs. Models are expected to keep developing while interfacing with Internet of Things (IoT) platforms as well as smart grids to enable automatic operation of every aspect of wind energy production. The core element for optimizing resource functions in wind energy systems driven by Artificial Intelligence stands as machine learning. The technology shows great potential to transform green energy by implementing intelligent forecasting techniques, adaptive control systems, and data-based decision-making because it creates vital infrastructure for sustainable, resilient, and efficient power systems.

REFERENCES

[1] M.B. Abdelghany, V. Mariani, D. Liuzza, O.R. Natale, and L. Glielmo, "A Unified Control Platform and Architecture for the Integration of Wind-Hydrogen Systems Into the Grid", *IEEE Transactions on Automation Science and Engineering,* vol. 21, no. 3, pp. 4042-4057, 2024. [http://dx.doi.org/10.1109/TASE.2023.3292029]

[2] Global Wind Energy Council (GWEC), Global Wind Report 2023, Brussels, Belgium, 2023. Available from: https://gwec.net/global-wind-report-2023/

[3] SS Moses, and J Thornburg, "Explainable AI for Renewable Energy Project Prefeasibility Analysis", In: *Proceedings of the 2025 IEEE PES 17th Asia-Pacific Power and Energy Engineering Conference (APPEEC)* Auckland, New Zealand, 2025, pp. 1-6. [http://dx.doi.org/10.1109/APPEEC66370.2025.11382138]

[4] M. Bou-Rabee, K.A. Lodi, M. Ali, M.F. Ansari, M. Tariq, and S.A. Sulaiman, "One-Month-Ahead Wind Speed Forecasting Using Hybrid AI Model for Coastal Locations", *IEEE Access,* vol. 8, pp. 198482-198493, 2020. [http://dx.doi.org/10.1109/ACCESS.2020.3028259]

[5] A. Hasan, E. Kandemir, D. Mordasov, and D. Trong Nguyen, "Agentic AI in Wind Energy Systems: Multi-Agent Architectures for Optimization and Resilience", *IEEE Access,* vol. 14, pp. 9935-9951, 2026. [http://dx.doi.org/10.1109/ACCESS.2026.3654529]

[6] M.M. Gulzar, A. Iqbal, D. Sibtain, and M. Khalid, "An Innovative Converterless Solar PV Control Strategy for a Grid Connected Hybrid PV/Wind/Fuel-Cell System Coupled With Battery Energy Storage", *IEEE Access,* vol. 11, pp. 23245-23259, 2023. [http://dx.doi.org/10.1109/ACCESS.2023.3252891]

[7] P. Biswas, A. Rashid, A.A. Masum, M.A.A. Nasim, A.S.M.A. Ferdous, K.D. Gupta, and A. Biswas, "An extensive and methodical review of smart grids for sustainable energy management-addressing

challenges with AI, renewable energy integration and leading-edge technologies", *IEEE Access,* vol. 12, pp. 31776-31804, 2024.
[http://dx.doi.org/10.1109/ACCESS.2025.3537651]

[8] A. Clifton, N. Zhang, J.B. McCaa, B.L. Haupt, and A.D. Freedman, "The value of probabilistic forecasts of wind power", *IEEE Trans. Sustain. Energy,* vol. 4, no. 2, pp. 339-348, 2013.
[http://dx.doi.org/10.1109/TSTE.2012.2218535]

[9] I. Kaur, and A. Ali, "A Complete Study on Machine Learning Algorithms for Medical Data Analysis", In: *Fog Computing for Intelligent Cloud IoT Systems,* C. Banerjee, A. Ghosh, R. Chakraborty, A.A. Elngar, Eds., Wiley-IEEE Press: Hoboken, NJ, USA, 2024, pp. 245-274.
[http://dx.doi.org/10.1002/9781394175345.ch7]

[10] D.F. Guedes Filho, R.L. Miranda, L.A. Correia, L. do ES Fernandes, R.C. Medrado, and M. De C. Filho, "Optimization and Integration Strategies for Hybrid Renewable Energy Systems in the Brazilian Power Grid: A Systematic Review", *IEEE Access,* vol. 13, pp. 84170-84187, 2025.
[http://dx.doi.org/10.1109/ACCESS.2025.3569199]

[11] N. Prajapati, and S. Paul, "Development of the Wind Map for Small-Scale Wind Energy Conversion System Using Advanced AI-Based Approach in the Indian State of Rajasthan", In: *Proceedings of the 2025 IEEE 15th International Conference on Power Electronics and Drive Systems (PEDS)*Penang, Malaysia, 2025, pp. 1-5.
[http://dx.doi.org/10.1109/PEDS63958.2025.11144885]

[12] A. Arunkumar, P. Pandiyarajan, S.B.L. Akshaya, A. Archana, and A. Archana, "Renewable Energy Consumption on Solar and Wind Energy Prediction Using Deep Learning and Generative AI", In: *Proceedings of the 2024 International Conference on IoT Based Control Networks and Intelligent Systems (ICICNIS)*Bengaluru, India, 2024, pp. 770-777.
[http://dx.doi.org/10.1109/ICICNIS64247.2024.10823172]

[13] Palak, and B. Koul, "MPPT-Driven Wind Energy System Under Dynamic Wind Conditions", In: *Proceedings of the 2025 International Conference on Electronics, AI and Computing (EAIC)*Jalandhar, India, 2025, pp. 1-8.
[http://dx.doi.org/10.1109/EAIC66483.2025.11101521]

[14] S. Mammadov, M. Çakallıoğlu, Y. Yeşilyurt, and P. Canbay, "Analysis of Sustainability in Hydroelectric, Wind, and Solar Energy Systems with Artificial Intelligence", In: *Proceedings of the 2025 7th International Congress on Human-Computer Interaction, Optimization and Robotic Applications (ICHORA)*Ankara, Turkiye, 2025, pp. 1-5.
[http://dx.doi.org/10.1109/ICHORA65333.2025.11017224]

[15] D. Marimekala, S. Krishnan, and J. Lamb, "Harnessing AI to Optimize Energy Consumption in Academic Institutions to Combat Climate Change", In: *Proceedings of the 2025 IEEE Integrated STEM Education Conference (ISEC)*Princeton, NJ, USA, 2025, pp. 1-7.
[http://dx.doi.org/10.1109/ISEC64801.2025.11147246]

[16] C.R. Beniugă, E.B. Bedrule, D-L. Machidon, O.C. Beniugă, and D.M. Istrate, "AI-Based Optimization of Turbine Selection for Wind Farms Using Python and Machine Learning", In: *Proceedings of the 2025 International Conference on Electromechanical and Energy Systems (SIELMEN)*Iasi, Romania, 2025, pp. 674-678.
[http://dx.doi.org/10.1109/SIELMEN67352.2025.11260854]

[17] C. Sasikala, P. Palpandian, G. Muthuram, S.S. Kumar, S.S. Karthik, and C.M. Raj, "AI based Compact Controller for Wind farms Economical Operation, Control and Monitoring", In: *Proceedings of the 2024 2nd International Conference on Self Sustainable Artificial Intelligence Systems (ICSSAS)*Erode, India, 2024, pp. 331-334.
[http://dx.doi.org/10.1109/ICSSAS64001.2024.10760947]

[18] L. X. Hao, Z. X. Yang, A. Q. Si, and Y. P. Zhu, "A voltage controller for wind turbine generator," In: *2013 10th IEEE International Conference on Networking, Sensing and Control (ICNSC)*, Evry, France, 2013, pp. 925–930.

[http://dx.doi.org/10.1109/ICNSC.2013.6548861]

[19] W. Ai, X. Minru, Z. Haixiao, T. Wang, Y. Wang, C. Meng, S. Liu, Y. Zheng, and Y. Hu, "Study on Transmission Planning of Combined Wind and Storage System", In: *Proceedings of the 2018 2nd IEEE Conference on Energy Internet and Energy System Integration (EI2)*Beijing, China, 2018, pp. 1-4.
[http://dx.doi.org/10.1109/EI2.2018.8582476]

[20] J. Smith, "DeepMind AI increases the value of wind energy," Google AI Blog, Feb. 26, 2019. Available from: https://www.deepmind.com/blog/deepmind-ai-increases-the-value-of-wind-energy

[21] F.K. Karim, D.S. Khafaga, M.M. Eid, S.K. Towfek, and H.K. Alkahtani, "A novel bio-inspired optimization algorithm design for wind power engineering applications time-series forecasting", *Biomimetics (Basel),* vol. 8, no. 3, p. 321, 2023.
[http://dx.doi.org/10.3390/biomimetics8030321]

[22] Available from: https://www.iunera.com/kraken/fabric/support-vector-machines-svm/

[23] F.M.A. Hadi, H.H. Aly, and T. Little, "Harmonics Forecasting of Wind and Solar Hybrid Model Driven by DFIG and PMSG Using ANN and ANFIS", *IEEE Access,* vol. 11, pp. 55413-55424, 2023.
[http://dx.doi.org/10.1109/ACCESS.2023.3253047]

[24] A. Meng, S. Chen, Z. Ou, W. Ding, H. Zhou, J. Fan, and H. Yin, "A hybrid deep learning architecture for wind power prediction based on bi-attention mechanism and crisscross optimization", *Energy,* vol. 238, p. 121795, 2022.
[http://dx.doi.org/10.1016/j.energy.2021.121795]

[25] H. Dong, and X. Zhao, "Wind-Farm Power Tracking *via* Preview-Based Robust Reinforcement Learning", *IEEE Trans. Industr. Inform.,* vol. 18, no. 3, pp. 1706-1715, 2022.
[http://dx.doi.org/10.1109/TII.2021.3093300]

[26] A.A. Firoozi, F. Hejazi, and A.A. Firoozi, "Advancing wind energy efficiency: A systematic review of aerodynamic optimization in wind turbine blade design", *Energies,* vol. 17, no. 12, p. 2919, 2024.
[http://dx.doi.org/10.3390/en17122919]

[27] K.I. Khan, and A. Nasir, "Application of artificial intelligence in solar and wind energy resources: a strategy to deal with environmental pollution", *Environ. Sci. Pollut. Res. Int.,* vol. 30, no. 24, pp. 64845-64859, 2023.
[http://dx.doi.org/10.1007/s11356-023-27038-6] [PMID: 37097570]

[28] J. Chatterjee, and N. Dethlefs, "Scientometric review of artificial intelligence for operations & maintenance of wind turbines: The past, present and future", *Renew. Sustain. Energy Rev.,* vol. 144, p. 111051, 2021.
[http://dx.doi.org/10.1016/j.rser.2021.111051]

[29] I. Kaur, and A. Ali, "A Complete Study on Machine Learning Algorithms for Medical Data Analysis", In: *Fog Computing for Intelligent Cloud IoT Systems,* C. Banerjee, A. Ghosh, R. Chakraborty, A.A. Elngar, Eds., Wiley: Hoboken, NJ, USA, 2024, pp. 151-182.

[30] E.G. Nabati, and K.D. Thoben, "Data driven decision making in planning the maintenance activities of off-shore wind energy", *Procedia CIRP,* vol. 59, pp. 160-165, 2017.
[http://dx.doi.org/10.1016/j.procir.2016.09.026]

[31] S. Dash, S. Chakravarty, N.C. Giri, and R. Khargotra, "Evaluating sustainable wind energy sources with multiple criteria decision-making (MCDM) techniques", *Comput. Electr. Eng.,* vol. 123, p. 110285, 2025.
[http://dx.doi.org/10.1016/j.compeleceng.2025.110285]

[32] I.U. Haq, A. Kumar, and P.S. Rathore, "Machine learning approaches for wind power forecasting: a comprehensive review", *Discov. Appl. Sci.,* vol. 7, no. 10, p. 1139, 2025.

[33] Y. Wang, R. Zou, F. Liu, L. Zhang, and Q. Liu, "A review of wind speed and wind power forecasting with deep neural networks", *Appl. Energy,* vol. 304, p. 117766, 2021.

[http://dx.doi.org/10.1016/j.apenergy.2021.117766]

[34] L. Gigoni, A. Betti, M. Tucci, and E. Crisostomi, "A scalable predictive maintenance model for detecting wind turbine component failures based on SCADA data", In: *2019 IEEE Power & Energy Society General Meeting (PESGM)*Atlanta, GA, USA, 2019, pp. 1-5. [http://dx.doi.org/10.48550/arXiv.1910.09808]

[35] I.U. Haq, and A.H. Rather, "Generative AI-Driven Frameworks for Intelligent Wind Energy Forecasting and Optimization", In: *Enhancing Renewable Energy Systems With Generative AI* IGI Global: Hershey, PA, USA, 2026, pp. 403-426.

[36] M. Arrieta-Prieto, and K.R. Schell, "Spatially transferable machine learning wind power prediction models: v−logit random forests", *Renewable Energy,* vol. 223, p. 120066, 2024. [http://dx.doi.org/10.1016/j.renene.2024.120066]

[37] I. Ul Haq, "Diagnosis of Neurological Disease Using Bioinspired Algorithms", In: *Bio-Inspired Optimization for Medical Data Mining,* S. Srivastava, A. Anand, A. Kumar, B. Saini, P.S. Rathore, Eds., Wiley-IEEE Press: Hoboken, NJ, USA, 2024.

[38] D. Sehgal, I. Kaur, V. Sharma, B. Gautam, A. Singh, and N. Kumar, "AI-Driven Early Diabetes Prediction", In: *Proceedings of the 2024 IEEE 3rd World Conference on Applied Intelligence and Computing (AIC)*Gwalior, India, 2024, pp. 40-48. [http://dx.doi.org/10.1109/AIC61668.2024.10730970]

[39] I. Kaur, and A. Ali, "An In-Depth Exploration of Machine Learning Algorithms and Performance Evaluation Approaches for Personalized Diabetes Prediction", In: *Proceedings of the 2024 International Conference on Emerging Innovations and Advanced Computing (INNOCOMP)*Sonipat, India, 2024, pp. 532-538.

[40] I.U. Haq, V. Nandal, and H. Uppal, "Blockchain Applications in Aviation: Securing Transactions, Streamlining Operations, and Improving Passenger Experience", In: *Blockchain in the Tourism Industry: A New Era of Secure and Transparent Travel Solutions* Springer Nature: Cham, Switzerland, 2025, pp. 131-154.

CHAPTER 4

The Integration of AI and Green Energy: Innovations and Operational Improvements

Shivani[1] and **Satinder Bal Gupta**[1,*]

[1] *Department of Computer Science & Engineering, Indira Gandhi University, Rewari, Haryana, India*

Abstract: AI has transformed many industries, including the renewable energy sector. AI has the ability to assess large volumes of data, foresee expected outcomes, and boost procedures, which makes it a useful tool for improving the efficacy and durability of green energy sources. The fusion of AI and green energy symbolizes a critical turning point in the search for a sustainable and reliable future. The merging of AI and energy can be considered an important step in improving system performance, longevity, and trustworthiness in the power production sector. AI is changing the way energy operators function by providing them with innovative facilities and different operational techniques in the field of green energy. AI provides many benefits, such as effective management of systems, prediction of energy output, fostering the robustness of the grid, and examining huge datasets. The incorporation of AI in the energy sector results in innovative advancements that help revolutionize the approaches used for energy transmission, storage, and production. This transformation results in an energy system that is robust and flexible, as the advanced system improves the effectiveness and performance of the energy systems. This amalgamation of advanced technology (AI) in the energy sector has taken the field of energy generation to new heights by improving reliability, effectiveness, and environmental sustainability. This transformation has resulted in the involvement of automation in the maintenance process, which has led to improvements in the effectiveness of green energy sources. This article is focused on demonstrating the various ways in which AI can enhance the performance of various operations, support eco-friendly energy initiatives, and boost economic growth in the energy sector. It also helps in providing valuable insights and future predictions to improve the energy sector. In addition to reshaping the energy sector, AI and energy are also laying the groundwork for an environment that is more robust, effective, and green. The scope of this revolutionary change expands beyond the boundaries of technological innovation. It also involves building an environment in which energy is more environmentally friendly, easier to obtain, and aligned with the overall goals of economic expansion and environmental preservation.

* **Corresponding author Satinder Bal Gupta:** Department of Computer Science & Engineering, Indira Gandhi University, Rewari, Haryana, India; E-mail: satinderbal@igu.ac.in

Sugandha Singh, Inam Ul Haq, Sachin Ahuja, Akib Mohi Ud Din Khanday & Neha Sharma (Eds.)

Keywords: Artificial Intelligence, Clean energy, Green energy, Renewable energy, Sustainability.

INTRODUCTION

Clean, environmentally friendly energy sources are more vital than ever as the world adopts both ecological consciousness and advancements in technology. The need for energy sources that offer high output, low emissions, and a reliable energy supply is growing throughout industries. The energy market needs to transform rapidly to achieve objectives such as significantly reducing emissions.

while offering reliable and adaptable energy to boost economic growth. Artificial Intelligence (AI) is essential to achieving this perspective because it provides formerly unavailable possibilities needed to revolutionize energy generation, management, and utilization. Advancement in green energy is being propelled by the use of AI technology [1]. This aggregation provides a lot of benefits, including the management of smart power systems, optimization of energy, and forecasting of maintenance. This integration results in the enhancement of profits, reduction of expenses, and the promotion of a more sustainable, environmentally friendly future. The advanced AI technology has created a revolution in the field of energy by re-shaping different processes related to the production and transmission of energy, as well as its consumption [2]. The energy sector is being reshaped by AI technology, as this advanced technology has propelled the sector towards a safe, sustainable, and economically efficient future. This shift can be noticed in areas such as smart grid management, predictions of energy management, and the safety of nuclear power plants. This technology has completely advanced the process of energy production and management [3]. AI provides a lot of benefits by strengthening the robustness of smart grids, improving the efficiency of energy, and facilitating the efficient allocation of resources using advanced Machine Learning and Big Data Analytics techniques. It is forecasted that by the year 2030, the amalgamation of intelligent grid technology and AI-powered green energy initiatives would be able to achieve the financial gain of approximately $1.3 trillion. Also, AI could help in reducing the greenhouse gas emissions by 5 to 10% worldwide, and this figure can be considered as analogous to the total annual emissions of the entire EU (European Union) [4]. This integration has created a shift towards the increase in usage of green energy, which can be considered the most significant advancement facilitated by AI technology in the green energy sector. AI can help in examining the consumption of energy, minimizing wastage, and managing smart grids by boosting the incorporation and oversight of different energy sources. Additionally, one of the main advantages of combining AI and the energy sector lies in the implementation of condition-based maintenance that helps in enhancing the longevity and robustness of different energy systems by foreseeing issues that can lead to system failures in the future. However, this amalgamation is accompanied by several challenges. Initially, implementing AI systems into operation and combining them with the present framework comes at a substantial initial cost. Some energy firms may find such costs expensive, especially smaller businesses with smaller financial resources. Second, the energy industry handles massive quantities of highly confidential data, such as operational specifics, client data, and infrastructure statistics. It is essential to guarantee the integrity of this data, and Artificial Intelligence systems need to be safeguarded from malicious attacks and incursions. Complying with safeguarding information laws, like the GDPR, introduces an additional dimension of complication. Additionally, there is a lack of skilled AI specialists who are aware of AI technology and the associated energy market. It is crucial to make training and education investments in order to overcome this knowledge gap, which may hamper the industry's acceptance and progression of AI technologies [5]. The energy market has earned great benefits from the integration of AI technology, which includes improvements in production, distribution, and consumption processes of energy. As AI technology is likely to advance in the future, it will also benefit the energy sector by enhancing the resilience of the grid, reducing emissions of greenhouse gases, and strengthening the efficiency of environmentally friendly energy sources. Soon, the rise in demand response management and smart electricity systems is expected, which will grant consumers with intelligent systems and thus help in controlling their energy consumption. Also, the use of predictive maintenance will help in fostering the reliability of machinery and reducing downtime. AI and green energy together will also help in resolving global warming issues by providing the facility of capturing and storing carbon. With the advancement of technology and the amalgamation of AI into energy systems, a more sustainable and cleaner energy environment is expected in the future. In order to create a sustainable and successful future, it is necessary to embrace the evolving relationship between Artificial Intelligence and the green energy sector, as AI includes a lot of innovations that can help in enabling machines to execute operations that need human intellect. AI technology is continually advancing and can be considered highly effective, as it can help in improving a lot of areas through machine learning and image identification. So, this advancement provides a lot of opportunities for innovation in various sectors. The green energy sector can be considered as one of the fields that can benefit from the advancement of this technology. Green energy is produced through natural resources such as sunlight, rainfall, wind, tides, ocean waves, and heat, which are sustainable and can be generated again over time. It is projected that by the year 2025, the worldwide green energy market will grow and reach a value of $2.15 trillion. The green energy sector serves an essential part in decreasing emissions of carbon dioxide, mitigating the adverse environmental impacts of conventional sources of energy, and establishing an environmentally friendly environment.

RELATED WORK

A lot of work has been done by several researchers in this area, which is analyzed in this section, as shown in Table **1**.

Table 1. Related work.

Reference	Objective	Conclusion
[6]	An in-depth examination of frequently utilized AI techniques in applications incorporating environmentally friendly sources of energy was presented. AI can be utilized for almost every form of energy source, which includes wind, solar, geothermal, water, ocean, bio, hydrogen, as well as hybrid, for architecture, optimization, forecasting, management, exchange, and supervision. The research aimed to showcase the AI approaches that can be used in the field of renewable energies.	Deep learning and ML techniques are essential for improving operations related to green energy and power production. The authors concluded that in the upcoming years, there is a high probability that AI techniques will advance.
[7]	The authors described the community-level services and challenges associated with Green AI. They explored the accuracy rates of algorithms based on a machine learning approach for diverse time frames.	The authors concluded that further study needs to be done at the community level, and different factors must be taken into account.
[8]	To investigate the most advanced Artificial Intelligence (AI) methods and resources for power regulation, maintenance, and administration of energy systems that are renewable, giving priority to solar power systems.	The Artificial Neural Network (ANN), Back Propagation Neural Network (BPNN), Adaptive Neuro-Fuzzy Inference System (ANFIS), and Genetic Algorithm (GA) were among the several theoretical and practical AI strategies that were often employed and shown to be trustworthy. As per the results of this study, these methods are often used in various kinds of solar applications. ANN, however, was the most efficient of all methods. The quick computation time, increased accuracy, and generalization capabilities of ANN over other modelling approaches are some of its unique benefits over its rivals. Compared to alternative modelling approaches, this would result in cost savings.

(Table 1) cont.....

Reference	Objective	Conclusion
[9]	To assess the applications of numerous AI approaches across a wide variety of clean energy techniques, such as wind, geothermal, energy storage, control of electricity, and photovoltaics.	AI algorithms and methods have shown potential in addressing the challenges associated with energy from geothermal sources. Several AI modelling techniques and their potential implications in generating electricity from sunlight have been examined. As they may generate a concise strategy for multi-variable problems, encompass less processing effort, and lack the need for having insight into internal factors, these models have shown themselves to be an effective substitute for classic physical simulation techniques.
[10]	This study was mainly aimed at highlighting the impact of AI in the domains of sustainable development and clean energy, conducting a systematic evaluation using the PRISMA methodology.	AI has emerged as an essential strategy for achieving major goals like SDG 7 (Affordable and Clean Energy) and SDG 13 (Climate Action), in the realms of clean energy and environmentally friendly development. Advanced techniques used for making predictions have improved the precision required in making predictions related to wind, speed, and exposure to sunlight and the production process of biofuel. The integration of advanced technology has fostered the efficacy and durability of different green energy sources. For instance, the implementation of machine learning algorithms has significantly improved the accuracy of forecasting trends related to wind and speed, which are critical for managing the production of wind power.
[11]	To analyze and assess the research related to Green and AI and environmentally friendly AI, utilizing a dual analytical model that merges BERTopic modeling and theme analysis. The main purpose of this method is to determine theme clusters as well as more intricate and continuously evolving topics in the concerned field.	The bibliometric analysis performed by the authors related to Green and eco-friendly AI determined three thematic clusters, and each of the clusters highlights a specific aspect of the evolution in this field. Responsible AI for eco-friendly development is the first and foremost cluster that is aimed at emphasizing the need to incorporate sustainable development principles into AI practices. Optimization represents the second cluster, which is aimed at making technological advancements so as to reduce the consumption of energy with the help of AI. Computational advancements using big data represent the third cluster that is aimed at incorporating AI and BDA (Big Data Analytics) in the energy sector, so as to highlight technological advancements and their consequences on the environment.

(Table 1) cont.....

Reference	Objective	Conclusion
[12]	A thorough examination of the present landscape of research on the integration of clean energy and AI, emphasizing significant methodologies, obstacles, and accomplishments.	The authors have discussed the role of different machine learning algorithms, neural networks and strategies related to optimization so as to highlight their impact on intricate datasets; enhance predictive accuracy and adaptive management of energy systems. The authors predicted that the integration of AI in the energy sector is optimistic in the coming future as more advancement in technology is expected to contribute in enhancing the efficiency and robustness of the energy system.

INTERSECTION OF AI AND GREEN ENERGY

The amalgamation of AI and green energy has created a revolution by making use of the ability of AI technology to boost the efficiency of various green energy sources. AI provides the facility of examining the data in real-time, which helps in enhancing the operational efficacy by foreseeing variations in demand for energy and optimizing the usage of energy. Advanced machine learning algorithms help in improving the performance of solar panels and wind turbines, which enables them to adjust to the changes and maximize the energy production [13]. This adjustment helps in combining energy sources with power grids, making the operations smooth by making use of different forecasting strategies and predictive modelling. AI has the ability to monitor various environmental conditions in real time including the assessment of whether trends, wind speed, and so on, which helps in improving the reliability of energy systems in changing environment conditions [14, 15]. Also, the predictive maintenance feature provided by AI helps in improving the sustainability, timely maintenance of the equipment, lowering the downtime, and improving the durability of various components. This amalgamation opens new doors of innovations for the energy sector by strengthening the reliability and efficiency of different operations needed in the energy sector. Thus, AI has helped in unlocking the capabilities of the green energy sector and making it more sustainable and environmentally friendly.

MARKET SIZE OF AI IN THE RENEWABLE ENERGY (GREEN ENERGY) SECTOR

The worldwide market for the adoption of AI technology in renewable energy is forecasted to have a value of USD 20.63 billion in 2025, with a presupposition to grow to approximately USD 158.76 billion by 2034, reflecting a compound annual growth rate (CAGR) of 25.65% during the period from 2025 to 2034, as shown in Fig. (**1**) [16].

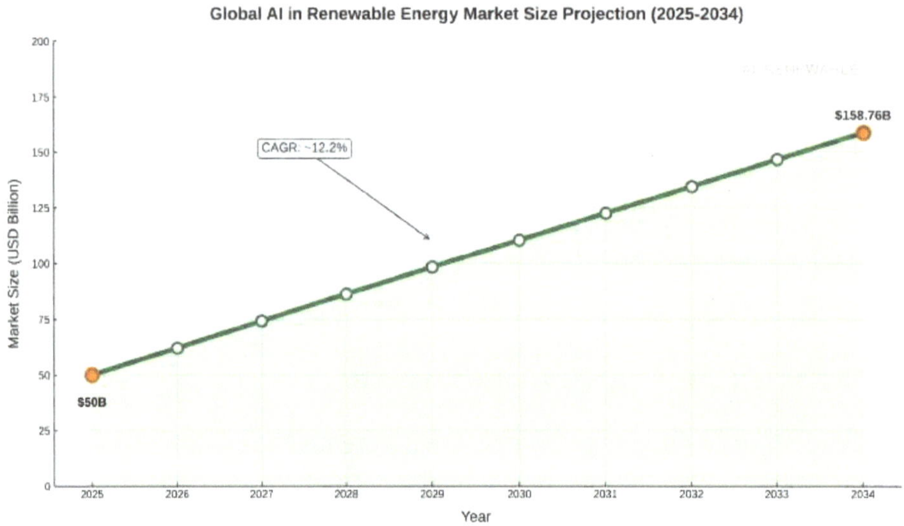

Fig. (1). Global market size of AI in the renewable energy sector.

The increasing demand for AI in the energy sector has driven the necessity for advanced energy systems and proactive fault detection measures to mitigate failures and reduce risks. AI facilitates accurate forecasting, effective supply management, and enhanced coordination.

Regions classified as developed and developing, including Europe, North America, and Asia-Pacific, are increasingly embracing AI technology within the renewable energy sector. These areas are playing a crucial role in the expansion of the global AI market in renewable energy throughout the forecast period. Furthermore, the presence of key industry stakeholders is further propelling the growth of AI in this sector on a global scale. Industry participants are implementing various strategies to strengthen their positions within the AI landscape in renewable energy. The Asia Pacific region held a remarkable position in the world market, commanding the largest market share of 49% in 2024, as shown in Fig. (**2**).

The market size for AI in renewable energy within this region was recorded at $7.93 billion in 2024, which is anticipated to increase to approximately $78.59 billion by 2034, and this growth shows an increase in compound annual growth rate (CAGR) of 25.78% from 2025 to 2034. The Asia Pacific region, particularly countries like India, faces challenges related to insufficient electricity supply. This increasing demand for electricity is a key factor propelling the growth of AI in the renewable energy market in the Asia Pacific [16].

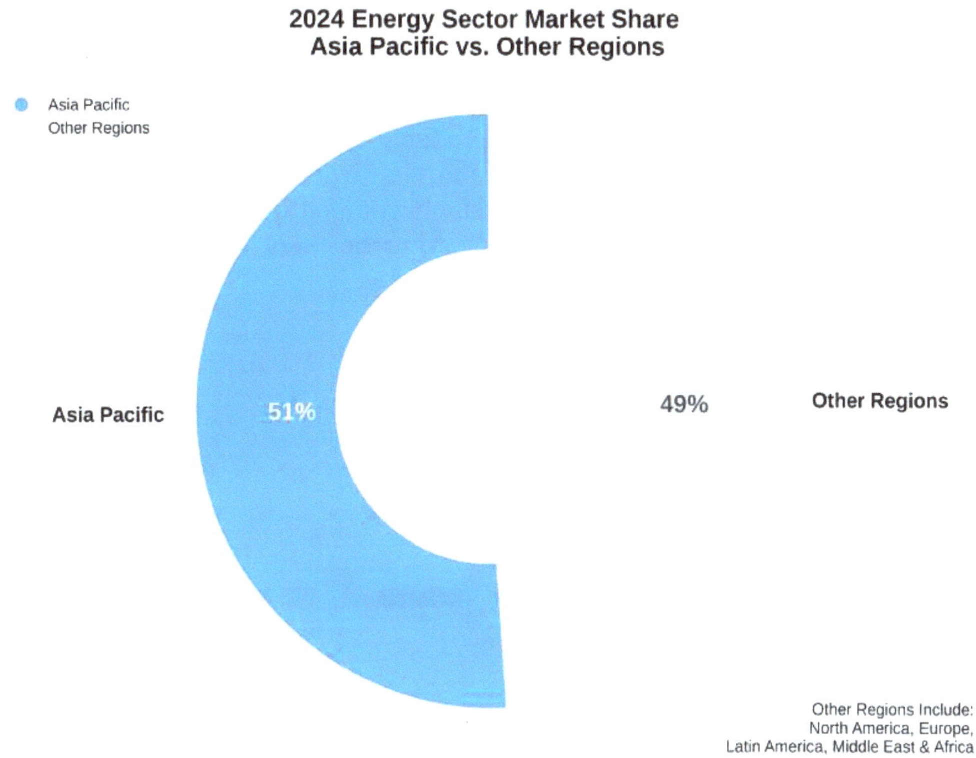

Fig. (2). Market share of Asia-Pacific in the energy sector in 2024.

North America is forecasted to increase significantly and experience a significant growth rate throughout the forecast period. It is anticipated to be the region with the highest growth rate during this timeframe. The advancement of AI within the renewable energy market in North America is primarily fueled by the increasing implementation and acceptance of advanced technologies and solutions within the energy sector. Additionally, the ongoing transformation and digitalization of the energy industry further contribute to the expansion of AI in this market. Furthermore, the amalgamation of Artificial Intelligence into smart home solutions is also facilitating the growth of AI technology in the clean energy sector across North America.

The growing emphasis on clean and sustainable energy is driving the incorporation of AI within the renewable energy sector. This development is anticipated to benefit the environment by mitigating the release of harmful gases associated with carbon and greenhouse gas emissions.

AI ADOPTION ADVANTAGES

The intelligent incorporation of AI into the energy market is a step towards a more productive, robust, and environmentally friendly green energy framework [17 - 22]. The deployment of AI in the production of energy, management, and dissemination has multiple benefits and puts leaders of companies in an advantageous position to thrive in an extremely competitive sector. The following are the primary advantages of incorporating AI in the energy market (Fig. **3**):

Fig. (3). Advantages of adopting AI in the energy sector.

Enhancement in the Efficiency of Operations

AI-driven advancements boost the operational efficacy of energy systems. AI guarantees optimum efficiency through smart grid management, maintenance forecasting, and power generation optimization. This pursuit of effectiveness maintains economic and ecological responsibility by lowering expenditures on operations, boosting energy asset results, and fostering resilience.

Cost Effectiveness

When AI is utilized in the energy sector, it helps in providing opportunities that result in a reduction of costs. The use of AI helps in automating the process of production, reducing wastage, and lowering the downtime required for maintenance. It thus helps in reducing the overall costs. In addition, the operational expenses can further be reduced by making use of forecasting techniques provided by AI, which helps in enhancing the planning of resource utilization and allocation of resources.

Dependability and Safety Enhancement

AI helps in identifying and addressing the potential issues that can arise in the future so as to improve the reliability of the process of supplying energy. It helps in forecasting the failure of equipment so as to avoid outages and maintain a continuous supply of energy. AI helps in increasing the safety by keeping track of danger signals and providing protection related to infrastructure, staff, and surroundings.

Renewable Energy Harmonization

As the need for green energy is increasing, the incorporation of AI plays a crucial role in fulfilling this demand. AI helps in making predictions regarding the availability of green energy sources and helps in enhancing their incorporation into the energy systems. This improvement in the utilization of energy resources helps in reducing the dependency and usage of fossil fuels and curtails the emission of carbon.

Declared Decision-Making

AI helps these stakeholders to get insightful information about the energy sector and thus, improve their capabilities of making informed decisions. It helps in analyzing the massive quantity of datasets; it provides valuable information related to market trends, behavior of consumers, and efficiency of operations. This provided knowledge helps in making decisions related to the investment strategy, change in operational process, and effectiveness in future planning so as to remain competitive in the future market.

Securing Forthcoming Viability

The adoption of AI is of great benefit for the energy sector as it provides innovative strategies to adapt to industrial needs. Thus, organizations should embrace using AI so as to obtain an environmentally friendly, digital, and efficient future.

OPPORTUNITIES FOR AI IN THE GREEN ENERGY SECTOR

AI can be considered a significant promoter within the green energy sector as the world faces an urgent need for energy resources that are environmentally friendly and sustainable. AI provides different opportunities that help in improving the durability and effectiveness of different energy sources by making use of machine learning models and algorithms [23 - 30]. AI has the power to make transformations for resolving issues faced by the green energy systems, including improvement in production and storage of electricity, forecasting failures, and enabling management of smart grids. This section is aimed at outlining various effects of AI that are beneficial for creating a transformation in the green energy sector.

Enhancing Production of Energy through AI

It is necessary to streamline the production of green energy for the purpose of creating an energy system that is environmentally sustainable. AI Technology incorporates machine learning algorithms that can adapt to the changing conditions of the environment. This strategy helps in resolving the intermittent nature of energy resources and increases the output, which promotes a more stable and eco-friendly energy system.

Fostering Efficiency of the Grid

Power Grid can be considered a critical component needed for the energy distribution system. The amalgamation of AI and smart grids helps in keeping track of the system in real-time and adaptive management of the electricity system as AI is an advanced technology, it helps in optimizing the process of distributing energy, minimizing the loss in transmission of energy, and ensuring smooth integration of multiple energy resources so as to enhance the efficiency of the grid.

Predictive Maintenance through AI

The main challenge faced by the green energy sector is related to the maintenance of equipment. The usage of AI in the energy systems helps in making predictions related to the failure of equipment, which helps in lowering the downtime, increasing the longevity of energy assets, and reducing the maintenance costs. This helps in increasing the long-term reliability of the energy sector.

CHALLENGES OF IMPLEMENTING AI IN THE GREEN ENERGY SECTOR

It cannot be denied that the incorporation of AI into the energy sector has created a great impact, but this integration faces a lot of challenges. Diffusion of AI into the energy systems requires considerable investments in technology and a proficient workforce [30 - 40]. One of the major challenges is to overcome the complexity of AI algorithms by providing high-quality data for training the algorithms. Issues regarding privacy and security of data, probable risk of algorithm bias, add more complexity to the adoption of AI technology in the concerned sector. This section is aimed at examining the various challenges that need to be addressed so that AI can be adopted in the field of energy completely, and a more environmentally friendly future can be promoted.

Data Availability and Quality for AI Applications in Green Energy

AI can be effectively implemented in the green energy sector only by focusing on the availability and quality of data. It is necessary to develop robust systems for collecting data of high quality to maintain the integrity of data. This helps in making accurate predictions, efficient management of energy systems, and informed decision-making, which improves the effectiveness of energy systems.

Ethical Considerations of Using AI in the Renewable Sector

Ethical considerations can be considered a significant area that needs to be focused on as a result of integration. Priority must be given to the openness, fairness, and accountability in the machine learning algorithms to develop trust among the involved stakeholders. The ethical AI standards need to be followed so as to ensure a balance between technological innovations and ethical responsibilities.

Addressing Regulatory Obstacles while incorporating AI and Green Energy

Amalgamating AI and green energy systems requires a complete understanding of complex regulatory frameworks. It is necessary to work with regulatory bodies to formulate guidelines that define the incorporation of AI technology in the green energy sector. Policymakers should be involved, as it is essential for maintaining a regulatory environment to boost innovation while ensuring the safety of the public interest during the integration process.

CASE STUDIES OF AI AND ML IN RENEWABLE ENERGY SYSTEMS

The impact created by AI and ML technologies in the energy sector can be highlighted through different practical case studies [40 - 45]. The different use cases showcased in this section highlight the effective amalgamation of technologies in different aspects of green energy, comprising improvement in the production of energy, optimization in the management of the grid, and predictive maintenance. Detailed analysis of these case studies provides informed insights into the practical applications, benefits, and challenges that are associated with the amalgamation of AI and ML in the green energy sector. This section is aimed at providing valuable examples that help in highlighting the usage of AI and ML in providing advanced energy solutions and thus, promoting a future that is more environmentally friendly.

Google's DeepMind and Wind Energy

The collaboration between the energy market and Google's DeepMind is aimed at optimizing the production of wind energy. The different machine learning algorithms are used to predict wind trends, which have improved the ability of DeepMind to accurately predict the wind energy. So, this innovation has resulted in higher energy generation and more efficient integration of wind power in the energy grid.

IBM's Watson and Solar Energy

IBM's Watson is useful in solar energy systems as it helps in enhancing the production of energy and different processes related to the maintenance of energy. It helps in providing valuable insights by examining data related to weather, performance of equipment, and different patterns related to the consumption of energy, which helps in enhancing the overall efficiency of solar panels and lowering the expenses of different operations.

Siemens Gamesa and Predictive Maintenance

Siemens Gamesa, a leading manufacturer of wind turbines, uses AI to provide the facility of predictive maintenance. It examines the data obtained from turbine sensors so as to forecast probable failures and schedule maintenance tasks on the basis of predictions. As a consequence, this method helps in reducing not only the downtime but also the costs of maintenance and thus, improves the efficiency of wind energy systems.

PowerScout

PowerScout, a renewable energy firm located in California, incorporates advanced technologies in the energy systems to improve the efficiency of operations. The AI technology incorporated in the systems helps in examining the data obtained from 100 billion data points, which represents approximately 45 million households, so as to predict the residences that are most likely to embrace solar energy. This helps in customizing the marketing strategy of different solar installation companies so as to increase their potential for generating revenue.

E.ON

E.ON, a leading German company, incorporates AI to handle variations in the production of renewable energy. AI helps E.ON in anticipating wind conditions at particular wind farm locations for the upcoming day and thus enables them to take proactive actions so as to address this variation in energy balance. AI also helps in improving the distribution and storage of energy and maintaining the balance between energy generation and consumption. The energy production and wind farms can also be enhanced in such a way that the system works efficiently with prevailing wind patterns.

HOW IS AI USED IN THE ENERGY SECTOR?

The entire energy industry is presently embracing Artificial Intelligence to significantly increase productivity and minimize expenses [45 - 52]. The energy industry may benefit greatly from AI in a variety of ways, from customer-oriented solutions like home automation to crucial scenarios like maintenance forecasting.

Smart Grids

Smart grids, which are electrical supply systems that rely on digital communications technological advances to determine and respond to regional distinctions in demand, can be administered with the assistance of Artificial Intelligence. First, by combining real-time and data from the past, AI systems can foresee consumption trends and may assist firms in more efficient distribution of resources. AI plays a substantial role in optimizing the process of resource allocation. It can improve the process of distributing electricity during unexpected surges in demand so as to ensure that power reaches the regions that greatly need it while reducing the risk of outages. Smart grids that make use of AI can determine faults or variations, including the failure of equipment, for service interruptions. AI algorithms can identify the exact location of these issues, which helps in the reduction of service interruptions, lowering of downtime, and improvement of grid efficiency.

Managing Demand and Supply

It plays a crucial role by boosting the efficiency of using energy and guaranteeing the reliability of the energy system. This approach is focused on making adjustments in the electricity usage habits of consumers, particularly in the commercial and industrial sectors. DRM implementation helps in balancing the supply and demand of energy during peak, facilitates load shifting so as to reduce the stress on the grid, and avoids costly modifications in the infrastructure. By facilitating instantaneous mode responses to alterations in energy demand, AI may additionally assist in determining a collaborative connection between energy suppliers and customers. AI can improve energy effectiveness, cut expenses, and facilitate the transition to clean sources of energy by anticipating and controlling variations in demand.

Predictive Maintenance

Energy firms can utilize AI to make accurate predictions when maintenance or breakdown of equipment is most likely to occur. ML helps in analyzing enormous quantities of data from an assortment of sources, such as usage data, climate trends, and previous maintenance reports, so as to anticipate possible issues before they occur. This method increases the entire dependability of energy infrastructure, reduces delays, and lowers the price of repairs.

Renewable Energy forecasting

AI is essential for predicting the manufacturing of green energy. Algorithms based on AI examine predictions for the weather, historical productivity data, and present situations for variable sources, such as solar and wind power. This makes it possible for energy providers to anticipate the quantity of green power that will be readily accessible, improving the demand-supply balance.

Energy Storage

AI maximizes the transmission and storage of green energy. AI algorithms decide when to preserve energy, when to discharge it, and how much to give away by taking into consideration several parameters, including demand, supply, pricing, and grid conditions. Green energy sources are sporadic. Energy storage makes it possible to preserve and utilize extra energy produced during times of high demand when these sources do not produce electricity. This minimizes the dependency of green energy sources on the climate and increases their reliability. Furthermore, storing energy is particularly crucial for important businesses where having an alternate source of energy could mean the difference between life and death, such as in hospitals, data centres, and emergency services.

Carbon Capture, Utilization, and Storage (CCUS)

Through the optimization of the process of capturing CO_2 from the environment or sources of emission, AI improves the effectiveness of CCUS systems. The most effective methods to use stored carbon, whether for safe long-term retention or manufacturing processes, can be determined by AI-driven systems. To slow down the impact of variations in climate and the reduction of emissions of greenhouse gases, this type of technology is essential.

Energy Trading

In energy trading, AI assesses intricate market conditions. It helps energy corporations make lucrative and well-informed trading choices by processing real-time data on availability, demand, and pricing fluctuations. AI also happens to be very good at handling risks and proactively assessing the unpredictable nature of the market. AI-powered algorithms for trading progress at unprecedented rates, completing numerous transactions in milliseconds. Energy portfolios are enhanced, market situations are recreated, sentiment is assessed, tasks are performed automatically, and it continuously adjusts to shifting market dynamics. Navigating the ever-changing marketplace for energy thus benefits greatly from AI's capacity to recognize emerging tendencies and patterns in enormous amounts of data. It recognizes market possibilities and dangers that ordinary traders might miss due to its remarkable pattern recognition skills.

Smart Homes and Buildings

AI is transforming homes and structures into energy-efficient ecological systems, which has a tremendously adverse effect on the search for energy efficiency. IoT devices and meters with sensors interact with AI to build flexible, smart ecosystems. These real-time, continuous power usage tracking devices enable AI to make data-driven judgments that maximize the consumption of energy. Imagine a future in which AI manages the functioning of cooling and heating systems. AI is capable of automatically adjusting the different temperature settings by considering different parameters like the preferences of customers, occupancy patterns, and prevailing environmental conditions. This marks a significant enhancement in living standards and a reduction in wastage of energy.

Oil and Gas Exploration

AI plays an essential part in the natural gas and oil exploration industry. AI can examine a massive amount of data with high accuracy, which enables it to determine the potential reserves of natural gas and oil, which was difficult to explore with the help of traditional techniques. Not only identification, but AI also helps in the evaluation of the feasibility of these reserves, directing exploration efforts towards the promising opportunities. This helps in enhancing the productivity and increasing the success rate of successful exploration, thereby saving both financial and material resources. The drilling operations are also influenced by the use of AI technology. Predictive algorithms are used in these operations that help in assessing different parameters such as geological formations, performance of drilling equipment, and surrounding conditions, so as to identify the probable risk and challenges. This prediction helps the drilling teams to improve safety measures, proactively resolve issues, and refine the process of drilling to make different operations safer and efficient.

Monitoring of Nuclear Power Plants

About 10% of the total energy generation is produced by nuclear power plants. While producing nuclear energy, the main concern is safety, and AI provides assurance for providing this safety. AI helps in continuously monitoring the different aspects of nuclear power plant operations. The data obtained from various sensors and devices is continuously analyzed by the AI system to detect the slightest deviations from accepted safety standards. AI makes use of advanced automatic repair simulations that help in identifying existing problems and predicting failures in equipment by examining the data concerning performance patterns, usage trends, and functional stressors. The strategies help the plant supervisors to take preventive actions and resolve issues before they increase and cause major incidents. Thus, AI provides help in maintaining high safety standards, supports accident prevention, and provides a steady supply of energy.

CONCLUDING REMARKS

The amalgamation of technologies such as AI and ML in green energy systems has created a revolution. The use of these innovative technologies has greatly enhanced the effectiveness and environmental sustainability of the energy facilities by streamlining the process of production and storage of energy and improving predictive maintenance and the robustness of the grid. The incorporation of these advancements has brought us closer to a future that gives priority to environmental conditions and green energy. The economic landscape for the usage of AI in the green energy sector is set to expand significantly, which marks remarkable growth and usage of AI in the industry. The use of AI in energy systems highlights the beginning of a sustainable and environmentally friendly future, as it provides great benefits and opportunities for the energy sector. STL Digital Solutions serves as an innovative example that shows the use of creative thinking, ethical considerations, and collaboration in creating a greener and environmentally friendly future. The fusion of AI and green energy is important for driving extraordinary advancements as we proceed through this journey of transformation to maintain a balance between technology and the environment.

REFERENCES

[1] G. Qingbin, P. Yanqing, and L. Kang, "The impact of Artificial Intelligence on energy environmental performance: Empirical evidence from cities in China", *Energy Economics,* vol. 141, 2025.

[2] H.-T. Yin, J. Wen, and C.-P. Chang, "Going green with Artificial Intelligence: The path of technological change towards the renewable energy transition," *Oeconomia Copernicana*, vol. 14, no. 4, pp. 1059–1095, Dec. 2023.

[http://dx.doi.org/10.24136/oc.2023.032]

[3] M. Shoaei, Y. Noorollahi, A. Hajinezhad, and S.F. Moosavian, "A review of the applications of Artificial Intelligence in renewable energy systems: An approach-based study", *Energy Conversion and Management,* vol. 306, 2024. [http://dx.doi.org/10.1016/j.enconman.2024.118207]

[4] I Ul Haq, IU Hassan, and HA Shah, "Machine Learning Techniques for Result Prediction of One Day International (ODI) Cricket Match", In: *Proceedings of the 2023 IEEE 8th International Conference on Convergence Systems and Innovative Technologies (I2CT)*, 2023.

[5] A. Ashfaq, M. Kamran, F. Rehman, N. Sarfaraz, H. U. Ilyas and H. H. Riaz, "Role of Artificial Intelligence in Renewable Energy and its Scope in Future", 2022 5th International Conference on Energy Conservation and Efficiency (ICECE), Lahore, Pakistan, pp. 1-6, 2022. [http://dx.doi.org/10.1109/ICECE54634.2022.9758957]

[6] A Azawii, S Al-Janabi, and OA Abdulteef, "Artificial Intelligence Techniques Applied on Renewable Energy Systems: A Review", *Proceedings of the International Conference on Computing, Communications and Networks (ICCCNet).,* Springer Nature: Singapore, pp. 297-308, 2022. [http://dx.doi.org/10.1007/978-981-19-0604-6_25]

[7] Y. Mehta, R. Xu, B. Lim, J. Wu, and J. Gao, "A Review for Green Energy Machine Learning and AI Services", *Energies,* vol. 16, no. 15, p. 5718, 2023. [http://dx.doi.org/10.3390/en16155718]

[8] J. T. Dellosa and E. C. Palconit, "Artificial Intelligence (AI) in Renewable Energy Systems: A Condensed Review of its Applications and Techniques," *IEEE International Conference on Environment and Electrical Engineering and 2021 IEEE Industrial and Commercial Power Systems Europe (EEEIC / I&CPS Europe)*, Bari, Italy, pp. 1-6, 2021. [http://dx.doi.org/10.1109/EEEIC/ICPSEurope51590.2021.9584587]

[9] IU Haq, AH Rather, and G Kaur, "A Comparative Analysis of Machine Learning Algorithms for the Early Prediction of Cardiovascular Disease", *Proceedings of the 2023 2nd International Conference on Edge Computing and Applications (ICECAA),* pp. 987-993, 2023. [http://dx.doi.org/10.1109/ICECAA58104.2023.10212214]

[10] R. Raman, S. Gunasekar, D. Kaliyaperumal, and P. Nedungadi, "Navigating the Nexus of Artificial Intelligence and Renewable Energy for the Advancement of Sustainable Development Goals", *Sustainability (Basel),* vol. 16, no. 21, p. 9144, 2024. [http://dx.doi.org/10.3390/su16219144]

[11] R. Raman, D. Pattnaik, H.H. Lathabai, C. Kumar, K. Govindan, and P. Nedungadi, "Green and sustainable AI research: an integrated thematic and topic modeling analysis", *J. Big Data,* vol. 11, no. 1, p. 55, 2024. [http://dx.doi.org/10.1186/s40537-024-00920-x]

[12] I Ul Haq, "Diagnosis of Neurological Disease Using Bioinspired Algorithms", In: *Bio-Inspired Optimization for Medical Data Mining,* S Srivastava, A Anand, A Kumar, B Saini, PS Rathore, Eds., Wiley-IEEE Press: Hoboken, NJ, USA, 2024. [http://dx.doi.org/10.1002/9781394214211.ch12]

[13] D. Navarra, "Integrating Artificial Intelligence and sustainable technologies in strategic renewable energy and Power-to-X projects: A review of global best practices, risks and future prospects", *Soc. Econ.,* vol. 45, no. 4, pp. 472-493, 2023. [http://dx.doi.org/10.1556/204.2023.00012]

[14] N. Rane, "Contribution of ChatGPT and other generative Artificial Intelligence (AI) in renewable and sustainable energy", *SSRN Electron J.*, Jan. 30, 202.. Available from: https://ssrn.com/abstract=4597674 [http://dx.doi.org/10.2139/ssrn.4597674]

[15] S.C. Necula, "Assessing the Potential of Artificial Intelligence in Advancing Clean Energy

Technologies in Europe: A Systematic Review", *Energies,* vol. 16, no. 22, p. 7633, 2023. [http://dx.doi.org/10.3390/en16227633]

[16] IU Haq, SZ Rufai, AH Rather, and HA Shah, "Transparency in Disease Diagnosis: Leveraging Interpretable Machine Learning in Healthcare", In: *Explainable Artificial Intelligence in the Healthcare Industry,* A Kumar, TA Kumar, P Das, C Sharma, AK Dubey, Eds., Wiley-IEEE Press: Hoboken, NJ, USA, 2025, pp. 112-135. [http://dx.doi.org/10.1002/9781394249312.ch6]

[17] A. Matharaarachchi, W. Mendis, K. Randunu, D. De Silva, G. Gamage, H. Moraliyage, N. Mills, and A. Jennings, "Optimizing Generative AI Chatbots for Net-Zero Emissions Energy Internet-of-Things Infrastructure", *Energies,* vol. 17, no. 8, p. 1935, 2024. [http://dx.doi.org/10.3390/en17081935]

[18] F. Yao, Z. Qin, X. Wang, M. Chen, A. Noor, S. Sharma, J. Singh, D. Kozak, and A. Hunjet, "RETRACTED: The evolution of renewable energy environments utilizing Artificial Intelligence to enhance energy efficiency and finance", *Heliyon,* vol. 9, no. 5, p. e16160, 2023. [http://dx.doi.org/10.1016/j.heliyon.2023.e16160] [PMID: 37234613]

[19] T Ahmad, D Zhang, C Huang, H Zhang, N Dai, Y Song, and H Chen, "Artificial intelligence in sustainable energy industry: Status Quo, challenges and opportunities", *Journal of Cleaner Production,* vol. 289, p. 125834, 2021.

[20] T. H. Nguyen, P. Paramasivam, V. H. Dong, H. C. Le, and D. C. Nguyen, "Harnessing a Better Future: Exploring AI and ML Applications in Renewable Energy", *JOIV : International Journal on Informatics Visualization,* vol. 8, no. 1, p. 55, 2024. [http://dx.doi.org/10.62527/joiv.8.1.2637]

[21] K.S. Arvind, and G.S. Madhuri, "An Energy Efficient Artificial Intelligence Based Innovation Detection for Complex Data Communication Model", In: *Proceedings of the 2023 International Conference on Distributed Computing, Electrical Circuits and Electronics (ICDCECE),,* pp. 1-7, 20232023. [http://dx.doi.org/10.1109/ICDCECE57866.2023.10151207]

[22] V. Bolón-Canedo, L. Morán-Fernández, B. Cancela, and A. Alonso-Betanzos, "A review of green Artificial Intelligence: Towards a more sustainable future", *Neurocomputing,* vol. 599, 2024. [http://dx.doi.org/10.1016/j.neucom.2024.128096]

[23] R. Vinuesa, H. Azizpour, I. Leite, M. Balaam, V. Dignum, S. Domisch, A. Felländer, S.D. Langhans, M. Tegmark, and F. Fuso Nerini, "The role of artificial intelligence in achieving the sustainable development goals", *Nat. Commun.,* vol. 11, no. 1, p. 233, 2020. [http://dx.doi.org/10.1038/s41467-019-14108-y] [PMID: 31911652]

[24] S. Aslam, H. Herodotou, S.M. Mohsin, N. Javaid, N. Ashraf, and S. Aslam, "A survey on deep learning methods for power load and renewable energy forecasting in smart microgrids", *Renew. Sustain. Energy Rev,* vol. 144, 2021.

[25] A. Tabbakh, L.A. Amin, M. Islam, G. M. I. Mahmud, I. K. Chowdhury and M. S. H. Mukta, “Towards sustainable AI: a comprehensive framework for Green AI”,. *Discov. Sustain.*, vol. 5, no. 408, 2024. [http://dx.doi.org/10.1007/s43621-024-00641-4]

[26] K. Qiu and K. Zhao, “The integration of green energy and Artificial Intelligence in next-generation energy supply chain: An analysis of economic, social, and environmental impacts”, *Sustain. Energy Technol. Assess.*, vol. 64, 2024. [http://dx.doi.org/10.1016/j.seta.2024.103660]

[27] N. Uriarte-Gallastegi, G. Arana-Landín, B. Landeta-Manzano, and I. Laskurain-Iturbe, "The Role of AI in Improving Environmental Sustainability: A Focus on Energy Management", *Energies,* vol. 17, no. 3, p. 649, 2024. [http://dx.doi.org/10.3390/en17030649]

[28] A.H. Rather, and I.U. Haq, "Intelligent Framework for Early Prediction of Diabetic Retinopathy: A

Deep Learning Approach", In: *Proceedings of the International Conference on Paradigms of Communication, Computing and Data Analytics (PCCDA),* pp. 377-388, 20232023. [http://dx.doi.org/10.1007/978-981-99-4626-6_32]

[29] J. Li, M.S. Herdem, J. Nathwani, and J.Z. Wen, "Methods and applications for Artificial Intelligence, Big Data, Internet of Things, and Blockchain in smart energy management", *Energy and AI,* vol. 11, p. 100208, 2023.

[30] P. Boza, and T. Evgeniou, "Artificial Intelligence to support the integration of variable renewable energy sources to the power system", *Appl Energy,* vol. 290, 2021. [http://dx.doi.org/10.1016/j.apenergy.2021.116754]

[31] D. Syed, A. Zainab, A. Ghrayeb, S.S. Refaat, H. Abu-Rub, and O. Bouhali, "Smart grid big data analytics: survey of technologies, techniques, and applications", *IEEE Access,* vol. 9, pp. 59564-59585, 2021. [http://dx.doi.org/10.1109/ACCESS.2020.3041178]

[32] D. Jayasutha, R. V. Prabhu, S. M. Sassirekha, K. Maithili, P. Girija and R. S. Subramanian, Exploring the Intersection of Sustainable Energy, AI, and Ethical Considerations", In: *AI Applications for Clean Energy and Sustainability*, B. Singh, C. Kaunert, K. Vig, and B. A. Riswandi, Eds. Hershey, PA, USA: IGI Global, 2024, pp. 215–234.

[33] I. Daniela, "Artificial Intelligence and Renewable Energy Utilization", In: *Proceedings of the International Conference on Business Excellence, Sciendo,* vol. 18, pp. 2776-2783, 2024. [http://dx.doi.org/10.2478/picbe-2024-0231]

[34] M. Khalil, and S.A. Sheikh, "Advancing Green Energy Integration in Power Systems for Enhanced Sustainability: A Review", *IEEE Access,* vol. 12, pp. 151669-151692, 2024. [http://dx.doi.org/10.1109/ACCESS.2024.3472843]

[35] Adetomi Adewumi, C.E. Okoli, F.O. Usman, K.A. Olu-lawal, and O.T. Soyombo, "Reviewing the impact of AI on renewable energy efficiency and management", *Int. J. Sci. Res. Archive,* vol. 11, no. 1, pp. 1518-1527, 2024. [http://dx.doi.org/10.30574/ijsra.2024.11.1.0245]

[36] J. Bhattacharjee and S. Roy, “Review on green resources and AI for biogenic solar power”, *Energy Storage Convers.*, vol. 2, no. 1, 2024. [http://dx.doi.org/10.59400/esc.v2i1.457]

[37] A.M.A. Shareef, S. Seçkiner, B. Eid, and H. Abumeteir, "Integration of blockchain with Artificial Intelligence technologies in the energy sector: a systematic review", *Front. Energy Res,* vol. 12, 2024.

[38] M. Hamid, and A. Ganne, "Artificial Intelligence in energy markets and power systems", *J. Eng. Technol. Manage.,* vol. 5, pp. 2582-5208, 2023.

[39] Q. Wang, Y. Li, and R. Li, "Ecological footprints, carbon emissions, and energy transitions: the impact of Artificial Intelligence (AI)", *Humanit Soc Sci Commun,* vol. 11, 2024.

[40] T. T. Le, J. C. Priya, H. C. Le, N. V. L. Le, M. T. Duong, and D. N. Cao, "Harnessing Artificial Intelligence for data-driven energy predictive analytics: A systematic survey towards enhancing sustainability", *Int. J. Renew. Energy Dev,* vol. 13, no. 2, pp. 270-293, 2024. [http://dx.doi.org/10.61435/ijred.2024.60119]

[41] I. Antonopoulos, V. Robu, B. Couraud, D. Kirli, S. Norbu, A. Kiprakis, D. Flynn, S. Elizondo-Gonzalez, and S. Wattam, "Artificial intelligence and machine learning approaches to energy demand-side response: A systematic review," *Renew. Sustain. Energy Rev.*, vol. 130, Art. no. 109899, 2020. [http://dx.doi.org/10.1016/j.rser.2020.109899]

[42] R. El-Araby, "Biofuel production: exploring renewable energy solutions for a greener future", *Biotechnol. Biofuels Bioprod,* vol. 17, no. 1, p. 129, 2024. [http://dx.doi.org/10.1186/s13068-024-02571-9] [PMID: 39407282]

[43] J. Chatterjee, and N. Dethlefs, "Facilitating a smoother transition to renewable energy with AI",

Patterns, vol. 3, no. 6, 2022..
[http://dx.doi.org/10.1016/j.patter.2022.100528]

[44] J.L. Holechek, H.M.E. Geli, M.N. Sawalhah, and R. Valdez, "A Global Assessment: Can Renewable Energy Replace Fossil Fuels by 2050?", *Sustainability,* vol. 14, 2022.

[45] K. Seetharaman, K. Moorthy, N. Patwa, Saravanan, and Y. Gupta, "Breaking barriers in deployment of renewable energy", *Heliyon,* vol. 5, no. 1, p. e01166, 2019.
[http://dx.doi.org/10.1016/j.heliyon.2019.e01166] [PMID: 30723834]

[46] T. Ahmad, H. Zhu, D. Zhang, R. Tariq, A. Bassam, F. Ullah, A.S. AlGhamdi, and S.S. Alshamrani, "Energetics Systems and Artificial Intelligence: Applications of industry 4.0", *Energy Rep.,* vol. 8, pp. 334-361, 2022.
[http://dx.doi.org/10.1016/j.egyr.2021.11.256]

[47] I. Sapre, "The Role of Artificial Intelligence in Enhancing Renewable Energy Efficiency: A Case Study on Solar and Wind Energy Optimization", *Intersect*, vol. 18, no. 1, 2024.

[48] A. Hussain, K. Kumar, and G. Ravikumar, "Bayesian-optimized bidirectional long-short-term memory network for wind power forecasting with uncertainty quantification", *Electr. Power Syst. Res.,* vol. 251, p. 112261, 2026.

[49] S. K. Srivastava, "Application of artificial intelligence in renewable energy", In: *2020 International Conference on Computational Performance Evaluation (ComPE)*, Shillong, India, 2020.

[50] B.N. Rao, M. Praveen, and R. Babu, "A review on the role of AI in optimizing renewable energy grid management", *Int. J. Sci. Res. Eng. Manag,* vol. 8, no. 11, 2023.

[51] N. S. Omerbegović, and D. Omerbegović, "Artificial intelligence in energy optimization and renewable energy system integration", In: *Artificial Intelligence in Chemical Engineering* Elsevier, 2026, pp. 471-498.18

[52] Z. Yao, Y. Lum, A. Johnston, L.M. Mejia-Mendoza, X. Zhou, Y. Wen, A. Aspuru-Guzik, E.H. Sargent, and Z.W. Seh, "Machine learning for a sustainable energy future", *Nat. Rev. Mater.,* vol. 8, no. 3, pp. 202-215, 2022.
[http://dx.doi.org/10.1038/s41578-022-00490-5] [PMID: 36277083]

CHAPTER 5

Wind Energy Prediction Utilizing Artificial Intelligence on Green Energy Transformation

Vikas Verma[1], **Tiyas Sarkar**[1,*] and **Manik Rakhra**[1]

[1] *School of Computer Science and Engineering, Lovely Professional University, Phagwara, Punjab, India*

Abstract: Recently, developments in renewable energy have revealed the overall significance of enhancing the use of Wind Energy (WE). There is a need to precisely forecast (WS) Wind Speed and (WP) Wind Power. Moreover, being able to predict them accurately is pivotal for energy balance and for informed decision-making regarding generation. Without a precise forecast, it is almost impossible for wind energy to reach its full potential and achieve better integration in the energy industry. Nowadays, there is an increasing improvement in forecasting techniques for WE, where AI (Artificial Intelligence) models have become the most reliable. These models excel over other methodologies for forecasting because they have a proven record of high accuracy and reliability. There are numerous research papers and quantitative experiments that have been conducted to test and validate AI models, which have demonstrated that AI models have successfully emerged as the best choice for predicting wind energy. With each additional generation of these systems, their predictive capabilities have continually improved. The objective of this paper is to provide a comprehensive review of the use of AI in WE forecasting, addressing factors such as energy production, cost, low-pressure wind areas, and many more. It assesses multiple Artificial Intelligence approaches and ranks them according to their capabilities and limitations.

Keywords: Artificial Intelligence, Energy optimization, Renewable energy, Wind speed forecasting, Wind power estimation, Wind energy prediction.

INTRODUCTION

The renewable energy sources are also gaining much acceptance in the power systems due to their economic viability. This makes it imperative to blend in sustainable energy sources into the power generation mix due to the ever-increasing global energy demands. Among these sources, wind energy has become a popular method of power generation because of its significant advanta-

* **Corresponding author Tiyas Sarkar:** School of Computer Science and Engineering, Lovely Professional University, Phagwara, Punjab, India; E-mail: fo.tiyasofficial11901657@gmail.com

Sugandha Singh, Inam Ul Haq, Sachin Ahuja, Akib Mohi Ud Din Khanday & Neha Sharma (Eds.)

ges; it is considered to be clean energy. The recent developments in wind energy technology have focused on optimizing resources in order to utilize wind energy to its full potential because it is one of the most widely available renewable energy resources and exists in abundant reserves in various regions of the world. In the last decade, Wind power has become one of the cheapest sources of energy on the globe, and has greatly expanded, which facilitates a more sustainable energy future. For example, China reached an impressive milestone of adding 71.67 GW in wind power capacity in 2020, which resulted in overall capacity leaping to 338 GW by the end of 2021. In the U.S, too, the efficiency of new wind power projects, with an average capacity factor of almost 40% attained in newly developed projects compared to older setups [1]. In 2023, the U.S. wind power capacity further increased by 13.4 GW, bringing the total capacity just below 136 GW by the end of the year. Some countries, like Brazil, India, Spain, Germany, and Italy, are leading the way in wind energy development and making many contributions toward its global development. As of 2021, the total wind power capacity in the world had increased to 839 GW. It is also noteworthy that China and the U.S. accounted for the largest share of the onshore wind energy market, as they together installed over 60% of the total new capacity for that year [2].

There are many issues to be addressed when integrating wind energy, especially at scale, into the Special Power Grid. The most troublesome aspect of Wind Energy is that it is highly erratic and varies greatly. This makes it extremely difficult to maintain balance between electricity “Demand and Supply”. Because of that, wind energy is often curtailed and is unable to be fully utilized. The solution to the main problem is to improve the wind power generation techniques as well as the forecasting techniques. With accurate wind predictions, better grid and turbine management can be achieved. This ultimately leads to smoother and safer functioning of the energy system as a whole. Additionally, wind energy revenue can increase with accurate forecasting. By accurately forecasting, the potential wind energy that the grid is able to consume can be predicted. This makes it easier to integrate wind energy and keep the system reliable. Even with the rapid development of wind turbines and solar panels across the globe, multiple issues persist. Changes in wind speed and direction make wind turbine energy output very difficult to manage. This has made predicting wind power generation extremely important for successful deployment and operation. Efficient use of wind energy remains a daunting challenge, as its intermittent nature leads to grid instability. In a bid to gather accurate information regarding wind conditions, researchers are developing artificial intelligence-based forecasting techniques. These efforts aim at improving scheduling and energy management decisions by optimizing the utilization of wind resources. Over the last couple of years, the investment in the development of artificial intelligence, *via* deep learning and machine learning methods, has commanded the most attention in Wind Forecasting.

However, earlier reviews have focused on classification problems in wind forecasting, which is only a small portion of what is dealt with in the discipline. This research aims to bridge the existing gap by providing an extensive review of wind prediction methodologies, with a focus on machine learning, deep learning, and hybrid techniques. It seeks to assist researchers in developing innovative and efficient wind forecasting systems. The chapter starts by examining contemporary studies on statistical methods for predicting wind speed (WS) and wind power (WP). The concepts and fundamental bases for improved wind energy predictors are detailed in Section 1. Section 2 concentrates on the sources and datasets used in forecasting, and Section 3 elaborates on the techniques that utilize artificial intelligence for wind energy predictors. Finally, Section 4 outlines some recommendations for future research in this field, followed by Section 5, where an overview of the results of the research is provided. As shown in Fig. (**1**), various Artificial Intelligence methods can be categorized into supervised learning, unsupervised learning, reinforcement learning, and hybrid approaches.

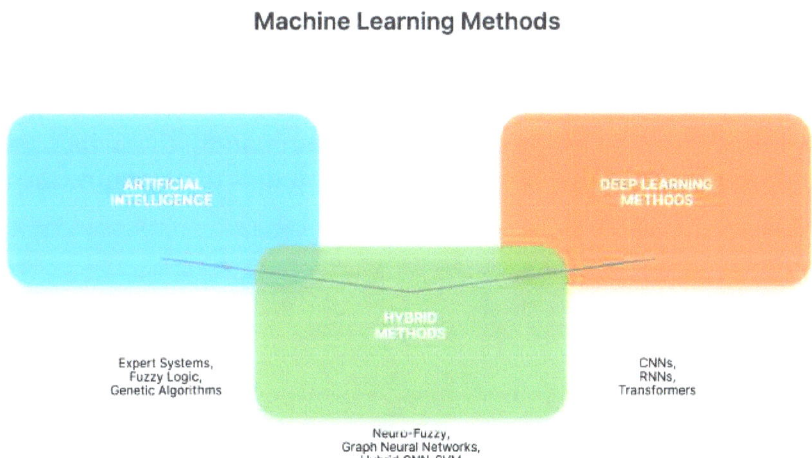

Fig. (1). An illustration of different methods of Artificial Intelligence.

OVERALL DESCRIPTION OF PREDICTING WIND ENERGY

Wind energy (WE) is the result of the movement of airstreams in the atmosphere due to solar-powered wind. It can be generated because of a heat imbalance that is caused by the configuration of the Earth's surface, contours, and the rotation of the Earth. As wind passes over the rotor blades of a wind turbine, which is specially designed to make a difference in air pressure, they experience aerodynamic forces in the form of lift and drag. Once the lift exceeds the drag, the rotor blades rotate, and this motion powers the generator for electricity production. Energy can be defined in terms of wind energy or power, just with the right formula like eq. (**1**):

$$P = \frac{1}{2}\,pAV^{3} \qquad \textbf{(1)}$$

Wind Turbine Power Output is identified by WP, and it depends on the power of the air (ρ), the area that is covered by the turbine's blades (A), and the factors of wind power (v). WP has a complicated non-linear function depending on wind speed (WS), and so it becomes very important to predict the WS to maximize power generation. It has been established that increasing the speed of the wind, energy forecasts by 10% could give a remarkable boost of 30% in overall energy forecasts by 10%.

So far, no common rule has been established for the foresight horizon, which is of particular use to them because it varies a lot. These periods are usually divided into long, medium, short or very short term as, in the Fig. (**2**), it is shown. Utilizing smart meters and other techniques, such as data compilation on the wind strengths for a certain wind farm, cuts across continents. A lot of work in the area has been done in direct power estimation for the wind turbines. Importantly, more than half of these studies add weather variables like temperature and wind speed in addition to the historical power data to models to improve the score. When a wind farm is completely new and does not have power data available, it is possible to calculate the energy output by estimating the wind speed. This technique is also used in preliminary feasibility studies to determine the capability of a new site for a wind farm.

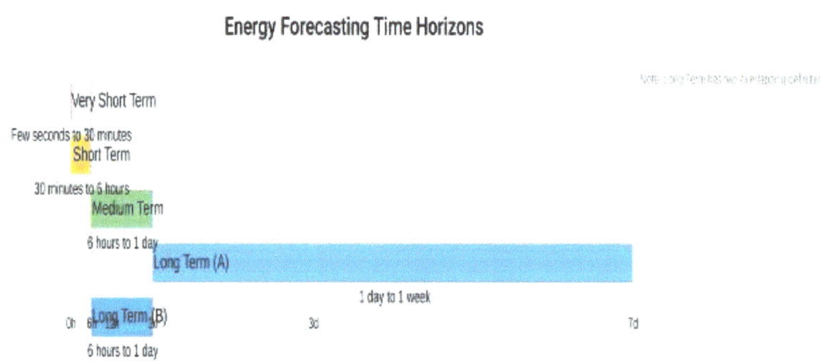

Fig. (2). To illustrate the Prediction of the Wind time series.

Very Short-Term: This time frame is used for tasks such as managing market issues in the power sector, ensuring compliance with regulations, grid control, controlling wind turbines, power quality analysis, load balancing, and supervising distribution systems.

Short Term: These are focused on power balancing, market load revisions, and triaging enforcement and operational-level security in the electricity market.

Medium-Term: These are applied for load shifts, energy dispatch, generator maintenance and operation, wind farm energy storage integration, congestion control, construction planning, energy and reserve procurement, and system planning.

Long-Term: These involve wind energy prospect assessment, wind farm construction and planning, economic assessment for operations, annual maintenance scheduling, conventional power plant management, routine operational supervision, feasibility studies of new wind farms, operational strategy formulation, energy market trading, and multi-source power plant optimization.

DATASET DESCRIPTION OF WIND ENERGY PREDICTION

The evaluation models in this work were created and tested through a variety of datasets from different areas. The formulation of power generation is done by using some specified equations. This method is indirect and provides more freedom. For instance, the predictions of wind speeds can be modified since the power output is a function of turbine parameters. Forecasting wind speed is less complicated than estimating wind power, because the latter is constrained by the parameters of the turbines [10]. But the accuracy of the indirect prediction methods can be enhanced with other meteorological variables in addition to the speed of the wind. The use of such meteorological data can reduce the impact of data volatility on the model and increase its performance in unstable environments.

PREDICTING WIND ENERGY USING ARTIFICIAL INTELLIGENCE

Given the great political importance of wind energy, Artificial Intelligence forecasting models are crucial for evaluating future wind energy generation by leveraging extensive historical datasets such as wind speed and direction. These models enable more reliable, acceptable predictions. Several techniques, like machine learning, deep learning, and hybrid methods, are used to improve the effectiveness of prediction. These methods are specially designed to work with the most complicated datasets and give appropriate portfolio forecasts. The primary purpose of Artificial Intelligence in Wind Energy forecasting is to optimize energy oversight. With productivity over the years, the importance of sustainable resources has put a significant burden on society and economic growth. With the implementation of AI in Wind Energy prediction, the adoption of wind power technologies increases as the predictions become more reliable, which leads to economic advantages. Fig. (**3**) shows AI workflow diagram where ML, DL, and hybrid algorithms are compared visually.

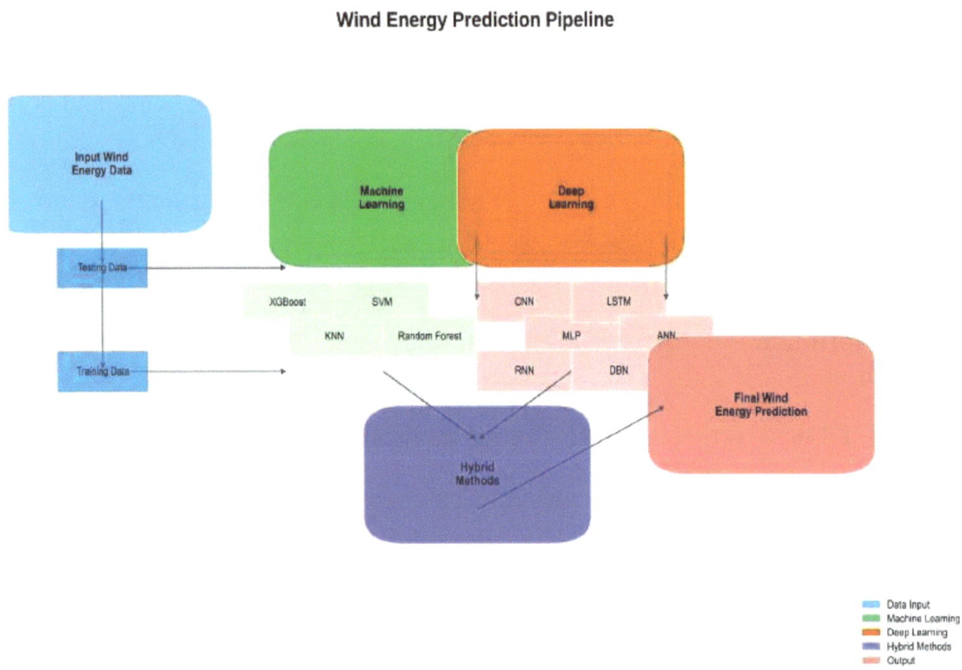

Fig. (3). To illustration of different models in Artificial Intelligence.

Wind Energy Prediction Using Machine Learning

Thanks to its ability to analyze massive datasets and extract valuable patterns, machine learning (ML) has become a preferred method for predicting wind energy generation. Some prevalent ML models include decision trees, random forests (RF), support vector regression (SVR), and gradient boosting (GB). However, a great deal of attention is placed on whether overfitting occurs. In the case of Wind Energy Prediction, overfitting can hamper the efficiency of the model on new data. Some changes or advancements in wind energy prediction with the aid of ML during the years 2022-2024 have been established and presented comprehensively.

For example, one of the studies pointed out the ability of RF to use wind speed (WS) and wind direction as input variables to predict wind power (WP). The advantage of RF is that it is powerful and provides accurate predictions with little tuning or optimization needed, so it was chosen for the task. Another benefit is that it is robust to meaningless features. Another study offered RF, GB, and XGBoost regression models as a solution to global and local wind energy and solar radiation forecasting problems. Their analysis showed that the ensemble approaches make SVR prediction for single wind farms more accurate, while the GB and XGBoost techniques showed high accuracy on forecasting wind energy at the regional level.

In another study, a method was proposed for long-term forecasting of wind power using daily wind speed data from five machine learning models. Support vector machines were found to be highly effective at predicting wind power values and could extend predictions to locations not included in the training dataset. This feature could enable the use of machine learning for feasibility testing at non-sited locations by applying models trained at existing locations. These studies were the first to propose a technique for wind power forecasting that integrates empirical mode decomposition EMD with Random Forest. This method decomposed the wind power series into intrinsic mode functions IMFs and residual components using EMD, and then trained the Random Forest with each component separately. The results from each component were combined to produce the final wind power prediction, thereby increasing the reliability of the results. For the forecasting of wind power for microgrid uses, another model was developed by blending fuzzy logic with model predictive control. The approach forecasted air density and WS, producing acceptable results within predetermined thresholds. Based on these fuzzy parameters, the models showed good predictive capabilities for WS and WP in the mentioned areas. In a different study, the focus was on evaluating the FP of short-term wind energy production in wind farms using five best ML optimization techniques –Random Forest, Gradient Boost, KNN, Decision Tree, and Extra Tree regression. The GB showed better performance than other techniques in this task, which leads to favourable expectations of the other methods. K-NN was shown to give the best prediction in the other study of wind power generation with various ML methods.

Like this, the algorithm TS_XGB was developed with the combination of spatial and temporal data mining for ultra-short-term wind power prediction. The algorithm was evaluated based on the 2018 records from the Xinjiang Datang Ruoqiang Wind Farm and demonstrated better results than Logistic Regression, ridge, SVM, or DT, marking its effectiveness. Lastly, additional research was conducted on the prediction of solar and wind energy generation using machine learning models, such as LR, ridge regression (RA), and ARIMA, for energy forecasting. ARIMA was the model that was most efficient and effective in the measurement of both energy sources, better than the rest. These improvements are a testament to the adaptability and accuracy of the ML algorithms in improving wind energy forecasting at different levels and varying conditions. The key parameters for wind energy prediction using machine learning are summarized in Table **1**.

Table 1. Parameters of wind energy prediction using machine learning.

Reference	Data (Input Level)	Approaches
[3]	Airflow velocity and bearing	Randomized Forest Regression (RFR)
[4]	Airflow velocity, energy, direction, and heat levels	Gradient Boosting Regressor (GBR), Extreme Gradient Boosting (XGB)
[5]	Wind speed measurement	LASSO, k-Nearest Neighbors (kNN), Extreme Gradient Boosting, Random Forest Regressor, Support Vector Regression (SVR)
[6]	Wind energy potential	Random Forest
[7]	Atmospheric density and airflow rate	Fuzzy Logic-Based Model
[8]	Airflow metrics and directional angles	Gradient Boosting Machine (GBM), k-Nearest Neighbors, Decision Trees, Extra Trees
[9]	Energy and airflow assessment	k-Nearest Neighbors
[10]	Velocity and directional flow	Extreme Gradient Boosting
[11]	Velocity and directional patterns	Autoregressive Integrated Moving Average (ARIMA)

Wind Energy Prediction Using Deep Learning

Deep learning models offer a powerful approach for handling large time-series data and are being used to predict wind energy generation.

The strength of these models lies in their ability to learn and capture underlying relationships and patterns from the data to inform decision-making. Their use, however, is limited in some wind energy forecasting applications due to their nature and the need for large amounts of data for training. For example, one of the studies reported the use of the Deep Belief Network (DBN) model for wind energy forecasting, where the Numerical Weather Prediction (NWP) data was processed with k-means clustering for faster computation of the NWP data. The performance of the DBN model was compared with the results from Backpropagation and Marlet Wavelet Neural Network (MWNN) Models. One more technique proposed selecting an optimal number of neurons in the hidden layer of ANN models to include 151 criteria for missing errors and improve convergence in Predicting Wind Speed. Also, the problem of 24-hour wind power forecasting was addressed using a hybrid deep learning model based on Convolutional Neural Networks (CNN) and Radial Basis Function Neural Networks (RBFNN), combining convolution, pooling, and uncertainty calculation techniques. The development of Stacked ELM (SELM) for enhanced memory optimization of the Extreme Learning Machine (ELM) strongly coincided with a multi-faceted approach involving deep learning. Additionally, the researchers proposed a CNN-based algorithm for wind speed forecasting, WindNet, which implemented a convolutional layer with fully connected components to analyze historical wind data. Furthermore, an advanced model that combined Wavelet Packet Decomposition (WPD), CNN, and CNN-LSTM to forecast wind speed was designed, separating high and low frequency components together with specific layers for precision prediction. Finally, another study hybridized support vector machines (SVM) with recurrent neural networks (RNN), long short-term memory (LSTM), and gated recurrent units (GRU) for better forecasting of wind speed, previously wavelet-transformed into smaller series. The primary parameters employed for wind energy prediction using deep learning approaches are presented in Table **2**.

Table 2. Parameters of wind energy prediction using deep learning.

Reference	Data (Input Level)	Approaches
[12]	Airflow rate, flow angle, total energy output database	Deep Belief Network (DBN)
[13]	Ambient temperature, flow angle, moisture level, airflow rate	Multi-Layer Perceptron (MLP)
[14]	Airflow rate	Convolutional Neural Network (CNN)
[15]	Airflow rate	Artificial Neural Network (ANN), Deep Neural Network (DNN), Extreme Learning Machine (ELM)
[16]	Airflow rate	Convolutional Neural Network (CNN)
[17]	Airflow rate	CNN combined with Long Short-Term Memory (CNN-LSTM)
[18]	Airflow rate	Layered LSTM-SM (Stacked Model)

Wind Energy Prediction Using Hybrid Model Approaches

The Hybrid Model for energy prediction capitalizes on the virtues of different machine learning (ML) and deep learning (DL) techniques to yield better results than single techniques. They also combine different models, such as Deep Neural Networks and classical Time Series models, to improve forecast accuracy and trustworthiness. The great merit of these hybrid methods is that they model the multidimensional interactions of the variables and find trends and patterns in the data. This makes them useful when predicting wind energy in highly variable and ever-changing situations. On the other hand, the application of hybrid models is more difficult compared to traditional methods because of the technical skills required as well as the increased computing costs. To address certain problems, new hybrid methods have been developed by researchers. For example, a stacking method combining XGBoost and Random Forest has been proposed to reduce overfitting, decrease wind speed dependency, and add important features with significant improvement in accuracy. The correlation-based filtering aided with VMD alongside LSSVM is used in conjunction with new wind speed forecasting algorithms. This helps to filter the features for wind speed. Other approaches include signal decomposition using ensemble empirical mode decomposition. The features are then optimized through genetic algorithms using recurrent neural networks (RNN) and units that use long short-term memory (LSTM). Others include the use of a hybrid approach with convolutional neural networks (CNN) and wavelet transforms (WT). The WT enables decomposition of wind energy data to provide higher prediction accuracy, while CNN captures different nonlinear features. Additionally, statistical analyses further improve prediction accuracy by modeling uncertainties in a probabilistic manner. Another model incorporating multiple techniques is MOGWO-WPD AdaBoost MRT-ORELM. This model is unique in its multi-step wind speed forecasting. Additionally, a WSTD-GRU method uses wavelets with soft threshold denoising to filter superfluous information while tuning for accurate forecasts. Other methods involve the use of weather simulation, optimization algorithms, fuzzy clustering, and models such as the SSOFC-Apriori-WRP, which use association rules, including Wind Speed Optimized Forecasting. This method performs wind speed forecasts for various segments and exploits their error pattern associations for better forecasting results. Finally, another hybrid deep learning method applies the ensemble empirical mode decomposition (EEMD) technique to raw data for use in a model. The resulting components are utilized to train a stack of restricted Boltzmann machines (RBM) to construct a Deep Boltzmann Machine (DBM) model, which enhances prediction accuracy. The parameters commonly applied in wind energy prediction using hybrid model approaches are listed in Table **3**.

Table 3. Parameters of wind energy prediction using hybrid model approaches.

Reference	Data (Input Level)	Approaches
[19]	Wind energy	Integration of XGBoost and Random Forest
[20]	Wind velocity	Least Squares Support Vector Machine (LSSVM)
[21]	Wind speed	Decomposition *via* EEMD with GA-optimized LSTM
[22]	Wind energy	Wavelet Decomposition paired with CNN
[23]	Wind forecasting	MOGWO-WPD combined with AdaBoost.MRT-ORELM
[24]	Wind speed	Wavelet Denoising integrated with GRU
[25]	Wind metrics	SSO-enhanced clustering and Apriori method (SSOFC-Apriori-WRP)
[26]	Wind velocity	EEMD-preprocessed data used in DBM

EVALUATION OF PERFORMANCE METRICS

In overcoming challenges in predicting intra- and inter-factors for energy systems, determining the model's dependability and the accuracy of its predictions are some of the baseline metrics. In this area, one needs to focus [27 - 35] on model optimization in a manner that improves their model while reducing errors. The effectiveness of the machine learning (ML) or any filtering model, such as deep learning (DL), is always judged by the criteria set in place. In this case, performance evaluation measures such as " RMSE, MAE, MSE, and R^2" may be used.

(RMSE)-Root Mean Square Error

In predictive modelling, a highly relevant metric while measuring accuracy is RMSE (Root Mean Squared Error). It is calculated using a set of outputs that have already been produced. RMSE measures how much the predictions deviate from the expected values. A smaller RMSE is advocated for how good a model is; an RMSE greater than zero signifies a prediction error.

(MSE)-Mean Squared Error

Mean Squared Error (MSE) is a metric that is usually termed as the most basic one while measuring the accuracy of a predictive model in a regression task. It determines the averages of the errors between predicted values and the set values. It places a larger emphasis on larger errors.

(MAE)-Mean Absolute Error

Mean Absolute Error (MAE) is a metric that nullifies all the negative aspects of algorithmic prediction, measuring the average of the absolute values of the errors in a set of predictions.

MAE is the average of the computed values of absolute differences between the predicted values and the sample outputs.

R Squared (R^2)

R-squared, also referred to as 'coefficient of determination,' is a metric that shows the precision of the predicted values from a regression model with respect to actual values. It provides an estimate of how much variance in the dependent variable is explained by the independent variable.

DISCUSSIONS AND OUTCOME OF FUTURE

The improvements in wind energy (WE) forecasting have resulted in a great number of studies claiming that the application of hybrid models, involving many techniques, is more accurate than non-hybrid models [38]. The hybrid approaches are widely used in modern studies. Their results are summarized in Table **4**. As some researchers have already provided a thorough comparative analysis of these technologies, their insights are not reiterated in this paper. We bring forward some directions for further research on existing problems and gaps in WE forecasting. Great attention should be focused on recent changes to deep learning (DL) technology in general, and in transformer models in particular. Such models are good for working with data arranged in sequences, as they extract features from long sequences and allow parallel execution of sequence elements [39]. Some models, such as Deep Transformer [40] and Informer [41], have been developed and are expected to yield good results with wind energy WE prediction problems. At the same time, other models of neural networks are being tested, such as graph neural networks, which shall allow a deeper understanding of the dependency relations existing between the multivariate Wind Energy data to enhance the accuracy of predictions [42]. With the attention-based neural networks outperforming traditional RNNs, especially for internal temporal dependencies, multi-resolution techniques are being implemented to identify concealed patterns [43 - 45]. There is also some research going on with graph convolutional networks, which are supposed to reveal the spatial relationships of neighbouring sites [46]. In the future, probabilistic forecasting, which accommodates uncertainty, is expected to dominate the approaches used in wind energy forecasting, overtaking the deterministic methods. One method to model the wind energy data's conditional probability distribution is utilizing conditional GANs [47]. Notably, the wind energy (WE) models prepared by the academics have their real-world utility, but interpretation is something that is still evolving. In simple terms, one understands interpretable models as those from which one can get relevant information when a certain problem is solved [48]. In WE prediction, interpretable deep learning (DL) models offer explanations with every forecast, which assists in building trustworthy and robust models while increasing the overall safety of the system. There is a need for more research in this area so that the wind energy industry will be effectively managed and risks reduced. The key parameters reported in existing research on wind energy prediction are summarized in Table **4**.

Table 4. Parameters of research on wind energy prediction.

Reference	Key Parameter Contribution
[36]	A refined version of the EMD-based method, known as EEMD, utilizes advanced techniques for calculating intrinsic mode functions and error metrics.
[37]	The domain requires an in-depth evaluation from various perspectives, highlighting emerging research directions and challenges.
[38]	Key aspects in the domain involve state-of-the-art forecasting methods, their overall accuracy, and strategies for enhancing performance.
[39]	The application of combined forecasting approaches and the maintenance of a current, comprehensive annotated bibliography are critical considerations.
[40]	A comparison of eight multi-step forecasting strategies reveals two enhanced prediction models.
[41]	Intelligent prediction models are categorized through two distinct approaches to highlight their strengths and weaknesses.
[42]	A new literature review can be conducted to track the development of wind energy research over the past three decades.
[43]	Applications of AI algorithms in the context of wind farms are explored.
[44]	37 evaluation criteria, 100 techniques, four classifications of wind data, three viewpoints, and 22 subcategories of forecasting methods are examined.

CONCLUSION

This review overviews the existing literature on different AI-based wind energy (WE) models. It classifies eight frequently used shallow and deep learning (DL) prediction methods within the scope of their advantages and disadvantages. The authors also elaborate on how each model compares in different forecasting situations and highlight their differences. The purpose of this review is to make a positive contribution to the AI-based WE forecasting techniques and aid the audience in grasping the various types of prediction techniques available. The article examines the use of Artificial Intelligence in Wind Energy forecasting. It covers a broad set of prediction algorithms that include machine learning (ML) and deep learning (DL) models such as Support Vector Regression (SVR) and Random Forest (RF), XGBoost, CNN, and RNN. To overcome the issues of single models, hybrid models have been developed that use stacking and ensemble techniques to unify several predictors. Every model comes with its unique set of pros and cons, depending on the goal of the forecast. For instance, RNN-based models specialize in identifying temporal patterns in a single times series data, while CNN-based models are good at finding relationships across multiple times series. The authors also suggest some directions for further research development in the area, such as applying optimization procedures for determining the model parameters and using more complex neural network structures. The focus of the study is primarily on the design and the structure of the predictive models.

ACKNOWLEDGEMENT

We extend our sincere gratitude to **Lovely Professional University (LPU)** for providing an exceptional research environment that fosters innovation and academic excellence. The university's unwavering support, access to state-of-t-e-art resources, and commitment to advancing knowledge have been instrumental in the successful completion of this research. We express our heartfelt appreciation to our mentors, colleagues, and the “Computer Science and Engineering” department at LPU for their invaluable guidance, constructive feedback, and encouragement throughout this study. Their insightful discussions and expertise have significantly contributed to refining our research. Furthermore, we acknowledge the immense support from the university’s research facilities, libraries, and technical staff, whose assistance has played a vital role in shaping our findings.

REFERENCES

[1] R. Saxena, "Artificial intelligence for renewable energy strategies and techniques", In: *Computer Vision and Machine Intelligence for Renewable Energy Systems,* A. K. Dubey, A. L. Srivastav, A. Kumar, U. C. Pati, F. P. García Márquez, V. García-Díaz, Eds., Academic Press (Elsevier): London, UK, 2025, pp. 17-39.

[2] M.S. Mauludin, M. Khairudin, R. Asnawi, W.A. Mustafa, and S.F. Toha, "The Advancement of Artificial Intelligence’s Application in Hybrid Solar and Wind Power Plant Optimization: A Study of the Literature", *J. Adv. Res. Appl. Sci. Eng. Technol.,* vol. 50, no. 2, pp. 279-293, 2024. [http://dx.doi.org/10.37934/araset.50.2.279293]

[3] I. Ul Haq, I. U. Hassan, and H. A. Shah, "Machine Learning Techniques for Result Prediction of One Day International (ODI) Cricket Match", In: *Proceedings of the 2023 IEEE 8th International Conference on Convergence Systems and Innovative Technologies (I2CT)*, 2023, pp. 1-5. [http://dx.doi.org/10.1109/I2CT57861.2023.10126241]

[4] T. Bashir, H. Wang, M. Tahir, and Y. Zhang, "Wind and solar power forecasting based on hybrid CNN-ABiLSTM, CNN-transformer-MLP models", *Renew. Energy,* vol. 239, p. 122055, 2025. [http://dx.doi.org/10.1016/j.renene.2024.122055]

[5] S. Adhikary, S. Sau, S. Debnath, and D. Ghosh, "Advancing bioenergy: leveraging artificial intelligence for efficient production and optimization", In: *Computer Vision and Machine Intelligence for Renewable Energy Systems* Elsevier, 2025, pp. 177-186.

[6] A.H. Rather, and I.U. Haq, "Intelligent Framework for Early Prediction of Diabetic Retinopathy: A Deep Learning Approach", In: *Proceedings of the International Conference on Paradigms of Communication, Computing and Data Analytics (PCCDA)*, A. Yadav, S. J. Nanda, and M. H. Lim, Eds. Singapore: Springer Nature, 2023, pp. 377–388. [http://dx.doi.org/10.1007/978-981-99-4626-6_32]

[7] D. Rout, "Solar energy generation and power prediction through computer vision and machine intelligence", In: *Computer Vision and Machine Intelligence for Renewable Energy Systems*, A. K. Dubey, A. L. Srivastav, A. Kumar, U. C. Pati, F. P. García Márquez, and V. García-Díaz, Eds. London, UK: Academic Press (Elsevier), 2025, pp. 103–123.

[8] R. K. Ahmad, S. A. Sulaiman, and Y. A. M. S. Anwar, "Artificial Intelligence in Biomass Energy: Trends, Tools, and Technologies," In: *Artificial Intelligence in Sustainable Energy and Environmental Management*, S. A. Sulaiman, Ed. Singapore: Springer Nature, 2026, pp. 21–45.

[9] D. Singh, A. Singh, M. Rakhra, T. Sarkar, G. S. Cheema, and A. Khamparia, "Predictions on the Future of Agriculture and Recent Developments in Agricultural Technology", In: *The Smart IoT Blueprint: Engineering a Connected Future,* F. Al-Turjman, Ed., Springer Nature: Cham, Switzerland, 2024, pp. 297-303.
[http://dx.doi.org/10.1007/978-3-031-63103-0_31]

[10] R. Mishra and D. K. Pandey, "Application of artificial intelligence in the development of circular bioeconomy," In: *Applications of Artificial Intelligence in Removal of Emerging Contaminants: Sustainable Approach for Environmental Clean-up and Circular Economy*, M. Sillanpää and R. T. Kapoor, Eds. Amsterdam, Netherlands: Elsevier, 2026, pp. 221–245.

[11] T. Sarkar, M. Rakhra, V. Sharma, A. Maan, A. Singh and G. S. Cheema, "Comparative study on Loan Approval Prediction Utilizing Machine Learning Approaches," In: Proceedings of the 2024 7th International Conference on Contemporary Computing and Informatics (IC3I), 2024, pp. 72–78.
[http://dx.doi.org/10.1109/IC3I61595.2024.10829014]

[12] H. Kaur, T. Sarkar, B. Kaur, A. Singh, G. S. Cheema and S. Das, "Using Machine Learning Techniques in Business Analytics: Lessons from Agile Project Management," 2024 7th International Conference on Contemporary Computing and Informatics (IC3I), Greater Noida, India, 2024, pp. 214-220.
[http://dx.doi.org/10.1109/IC3I61595.2024.10828636]

[13] T. Sarkar, M. Rakhra, V. Sharma, A. Singh, K. Jairath and A. Maan, "Comparing Traditional vs Agile Methods for Software Development Projects: A Case Study," 2024 7th International Conference on Contemporary Computing and Informatics (IC3I), Greater Noida, India, 2024, pp. 221-227.
[http://dx.doi.org/10.1109/IC3I61595.2024.10829321]

[14] S. Takkar, A. S. Rayat, V. Sharma, T. Sarkar, K. Jairath and A. Singh, "Problems and Optimal Methodologies Agile Project Management Solution in Publicly accessible Sector: A Comprehensive Evaluation of the Literature," 2024 7th International Conference on Contemporary Computing and Informatics (IC3I), Greater Noida, India, 2024, pp. 207-213.
[http://dx.doi.org/10.1109/IC3I61595.2024.10828832]

[15] S. Singh, T. Sarkar, M. Mangla, M. Rakhra, A. Singh and K. Jairath, "FinTech Edge: Utility Computing & Artificial Intelligence Technologies for Smart Financial Acquisition & Blockchain in the Financial Industries," 2024 7th International Conference on Contemporary Computing and Informatics (IC3I), Greater Noida, India, 2024, pp. 228-234.
[http://dx.doi.org/10.1109/IC3I61595.2024.10828908]

[16] N. Sharma, M. Sharma, A. Singhal, N. Fatema, V.K. Jadoun, H. Malik, and A. Afthanorhan, "A Spatiotemporal Feature Extraction Technique Using Superlet-CNN Fusion for Improved Motor Imagery Classification", *IEEE Access,* 2024.

[17] I. U. Haq, A. Kumar, and P. S. Rathore, "Machine learning approaches for wind power forecasting: a comprehensive review", *Discov Appl Sci,* vol. 7, p. 1139, 2025.
[http://dx.doi.org/10.1007/s42452-025-07675-x]

[18] A. Stetco, F. Dinmohammadi, X. Zhao, V. Robu, D. Flynn, M. Barnes, and G. Nenadic, "Machine learning methods for wind turbine condition monitoring: A review", *Renewable energy,* vol. 133, pp. 620-635, 2019.

[19] A. Alkesaiberi, F. Harrou, and Y. Sun, "Efficient wind power prediction using machine learning methods: A comparative study", *Energies,* vol. 15, no. 7, p. 2327, 2022.

[20] H. Liu, and Z. Zhang, "Development and trending of deep learning methods for wind power predictions", *Artificial Intelligence Review,* vol. 57, no. 5, p. 112, 2024.

[21] Y. C. Deng, X. H. Tang, Z. Y. Zhou, Y. Yang, and F. Niu, "Application of machine learning algorithms in wind power: a review", *Energy Sources, Part A: Recovery, Utilization, and Environmental Effects,* vol. 47, no. 1, pp. 4451-4471, 2025.

[22] A. Marvuglia, and A. Messineo, "Monitoring of wind farms' power curves using machine learning techniques", *Applied Energy,* vol. 98, pp. 574-583, 2012.

[23] T. Sarkar, M. Rakhra, V. Sharma, S. Takkar and K. Jairath, "Comparative Study of Object Recognition Utilizing Machine Learning Techniques," 2024 International Conference on Communication, Computer Sciences and Engineering (IC3SE), Gautam Buddha Nagar, India, 2024, pp. 726-731.
[http://dx.doi.org/10.1109/IC3SE62002.2024.10593475]

[24] A. K. Verma, Shruti and T. Sarkar, "Utilizing Imaging Steganographic Improvement using LSB & Image Decoder," 2024 International Conference on Communication, Computer Sciences and Engineering (IC3SE), Gautam Buddha Nagar, India, 2024, pp. 144-150.
[http://dx.doi.org/10.1109/IC3SE62002.2024.10593623]

[25] P. Chakrabarty, T. Sarkar, M. Rakhra, K. Jairath and V. Sharma, "Enhanced Data Security Framework Using Lightweight Cryptography and Multi-Level Encryption," 2024 International Conference on Communication, Computer Sciences and Engineering (IC3SE), Gautam Buddha Nagar, India, 2024, pp. 720-725,
[http://dx.doi.org/10.1109/IC3SE62002.2024.10593191]

[26] T. Sarkar, M. Rakhra, V. Sharma and A. Singh, "An Empirical Comparison of Machine Learning Techniques for Bank Loan Approval Prediction," 2024 International Conference on Communication, Computer Sciences and Engineering (IC3SE), Gautam Buddha Nagar, India, 2024, pp. 137-143.
[http://dx.doi.org/10.1109/IC3SE62002.2024.10593355]

[27] S. Talwani, M. Rakhra and T. Sarkar, "Perspectives on the Future of Agriculture and Recent Developments in Agricultural Technology," 2024 11th International Conference on Reliability, Infocom Technologies and Optimization (Trends and Future Directions) (ICRITO), Noida, India, 2024, pp. 1-6.
[http://dx.doi.org/10.1109/ICRITO61523.2024.10522413]

[28] T. Sarkar, B. Moharana, M. Rakhra and G. S. Cheema, "Comparative Analysis of Empirical Research on Agile Software Development Approaches," 2024 11th International Conference on Reliability, Infocom Technologies and Optimization (Trends and Future Directions) (ICRITO), Noida, India, 2024, pp. 1-6.
[http://dx.doi.org/10.1109/ICRITO61523.2024.10522134]

[29] A. K. Verma, T. Sarkar, M. Rakhra, V. K. Pandey and P. Chakrabarty, "Studying the Impact of Voice-Activated Chatbot's with Artificial Intelligence on Consumer Interaction in India," 2024 4th International Conference on Technological Advancements in Computational Sciences (ICTACS), Tashkent, Uzbekistan, 2024, pp. 1708-1713.
[http://dx.doi.org/10.1109/ICTACS62700.2024.10841074]

[30] B. Moharana, V. K. Singh, T. Sarkar, A. Hassan and V. K. Pandey, "Predicting Indian Master's and PhD Enrollment using Data Mining & Machine Learning Approaches," 2024 4th International Conference on Technological Advancements in Computational Sciences (ICTACS), Tashkent, Uzbekistan, 2024, pp. 1714-1720.
[http://dx.doi.org/10.1109/ICTACS62700.2024.10841260]

[31] T. Sarkar, M. Rakhra, P. Singh, P. K. Angra, G. S. Cheema and N. Mathur, "Comprehensive Review on Fog Computing: Obstacles, Feature Traits, Limitations, and Potential Leverages on Application," 2024 4th International Conference on Technological Advancements in Computational Sciences (ICTACS), Tashkent, Uzbekistan, 2024, pp. 404-410.
[http://dx.doi.org/10.1109/ICTACS62700.2024.10840690]

[32] T. Sarkar, A. Bhattacherjee and M. Mukhia, "Large Language Model for Practical Evaluation in Cloud

Computing with Applications of Real World Data," 2024 4th International Conference on Technological Advancements in Computational Sciences (ICTACS), Tashkent, Uzbekistan, 2024, pp. 1721-1729.
[http://dx.doi.org/10.1109/ICTACS62700.2024.10840404]

[33] D. B. Unsal, A. Aksoz, S. Oyucu, J. M. Guerrero, and M. Guler, "A Comparative study of AI methods on renewable energy prediction for smart grids: Case of Turkey", *Sustainability,* vol. 16, no. 7, p. 2894, 2024.
[http://dx.doi.org/10.3390/su16072894] [http://dx.doi.org/10.3390/su16072894]

[34] Ahmad Hamdan, and Sedat Sonko, "AI in renewable energy: A review of predictive maintenance and energy optimization", *Int. J. Sci. Res. Arch,* vol. 11, no. 1, pp. 718-729, 2024.
[http://dx.doi.org/10.30574/ijsra.2024.11.1.0112]

[35] H. N. N. Manuel, H. M. Kehinde, C. P. Agupugo, and A. C. N. Manuel, "The impact of AI on boosting renewable energy utilization and visual power plant efficiency in contemporary construction", *World J. Adv. Res. Rev.,* vol. 23, no. 2, pp. 1333-1348, 2024.
[http://dx.doi.org/10.30574/wjarr.2024.23.2.2359]

[36] M. Shoaei, Y. Noorollahi, A. Hajinezhad, and S.F. Moosavian, "A review of the applications of artificial intelligence in renewable energy systems: An approach-based study", *Energy Convers. Manage.,* vol. 306, p. 118207, 2024.
[http://dx.doi.org/10.1016/j.enconman.2024.118207]

[37] A. Masuda, H. Hebishima, and S. I. Inage, "A unified forecasting model for solar irradiation and wind speed using ground-based data", *Next Sustainability.,* vol. 7, p. 100225, 2026.

[38] V. Simankov, P. Buchatskiy, A. Kazak, S. Teploukhov, S. Onishchenko, K. Kuzmin, and P. Chetyrbok, "A Solar and wind energy evaluation methodology using Artificial Intelligence technologies", *Energies,* vol. 17, no. 1, p. 416, 2024.
[http://dx.doi.org/10.3390/en17020416] [http://dx.doi.org/10.3390/en17020416]

[39] S. Yousef, J. Eimontas, N. Striūgas, and M.A. Abdelnaby, "Co-pyrolysis of waste wind turbine blades and biomass and their kinetic analysis using artificial neural network", *J. Anal. Appl. Pyrolysis,* vol. 179, p. 106495, 2024.
[http://dx.doi.org/10.1016/j.jaap.2024.106495]

[40] I. O. Adewumi, R. Ibrahim, and K. Ibrahim, "Optimizing Cassava-to-Bioenergy Conversion Using Artificial Neural Networks: A Sustainable Pathway for Africa," *BioEnergy Research*, vol. 19, no. 1, p. 20, 2026.
[http://dx.doi.org/10.3390/en17020416]

[41] S. Yousef, J. Eimontas, N. Striūgas, and M.A. Abdelnaby, "Synthesis of value-added aromatic chemicals from catalytic pyrolysis of waste wind turbine blades and their kinetic analysis using artificial neural network", *J. Anal. Appl. Pyrolysis,* vol. 177, p. 106330, 2024.
[http://dx.doi.org/10.1016/j.jaap.2023.106330]

[42] V.N. Nguyen, W. Tarełko, P. Sharma, A.S. El-Shafay, W-H. Chen, P.Q.P. Nguyen, X.P. Nguyen, and A.T. Hoang, "Potential of explainable artificial intelligence in advancing renewable energy: challenges and prospects", *Energy Fuels,* vol. 38, no. 3, pp. 1692-1712, 2024.
[http://dx.doi.org/10.1021/acs.energyfuels.3c04343]

[43] Koroglu, Tahsin, and Elanur Ekici. "A Comparative Study on the Estimation of Wind Speed and Wind Power Density Using Statistical Distribution Approaches and Artificial Neural Network-Based Hybrid Techniques in Çanakkale, Türkiye." *Appl. Sci.*, vol. 14, no. 3, Art. no. 1267, 2024.
[http://dx.doi.org/10.3390/app14031267]

[44] U. Rusilowati, H. Rasmita Ngemba, R. Wahyudin Anugrah, A. Fitriani, and E. Dian Astuti, "Leveraging AI for superior efficiency in energy use and development of renewable resources such as solar energy, wind, and bioenergy", *Int. Trans. Artif. Intell. (ITALIC),* vol. 2, no. 2, pp. 114-120, 2024.
[http://dx.doi.org/10.33050/italic.v2i2.537]

[45] D. Song, G. Shen, C. Huang, Q. Huang, J. Yang, M. Dong, Y. H. Joo, and N. Duić, "Review on the Application of Artificial Intelligence Methods in the Control and Design of Offshore Wind Power Systems", *J. Mar. Sci. Eng.,* vol. 12, no. 3, p. 424, 2024.
[http://dx.doi.org/10.3390/jmse12030424] [http://dx.doi.org/10.3390/jmse12030424]

[46] Shedrack Onwusinkwue, Femi Osasona, and Ahmad Hamdan, Islam Ahmad Ibrahim Ahmad; Anthony Chigozie Anyanwu; Samuel Onimisi Dawodu; Ogugua Chimezie Obi, "Artificial intelligence (AI) in renewable energy: A review of predictive maintenance and energy optimization", *World J. Adv. Res. Rev,* vol. 21, no. 1, pp. 2487-2799, 2024.
[http://dx.doi.org/10.30574/wjarr.2024.21.1.0347]

[47] W.S. Udo, "Optimizing wind energy systems using machine learning for predictive maintenance and efficiency enhancement", *J. Renew. Energy Technol,* vol. 28, no. 3, pp. 312-330, 2024.

[48] A. Matharaarachchi, W. Mendis, K. Randunu, D. De Silva, G. Gamage, H. Moraliyage, N. Mills, and A. Jennings, "Optimizing Generative AI Chatbots for Net-Zero Emissions Energy Internet-of-Things Infrastructure", *Energies,* vol. 17, no. 8, p. 1935, 2024.
[http://dx.doi.org/10.3390/en17081935]

CHAPTER 6

Intelligent Winds: Transforming Wind Energy through Machine Learning

Sakshi[1], **Arushi Aggarwal**[1] and **Inam Ul Haq**[2,*]

[1] *Department of Computer Science and Engineering (CSE), Faculty of Engineering and Applied Technology (FEAT), Shree Guru Gobind Singh Tricentenary (SGT) University, Gurugram, Haryana, India*

[2] *School of Engineering and Technology (SET), CGC University, Mohali, Punjab, India*

Abstract: The application of machine learning (ML) in wind energy systems has notably enhanced efficiency, reliability, and performance. Wind energy, while sustainable, faces challenges due to its intermittent nature. Machine learning is revolutionizing the sector by enhancing predictions, optimizing operations, and improving grid integration, making wind power more efficient and reliable. This chapter discusses major ML applications, including wind power forecasting, fault diagnosis, turbine operation optimization, and energy management. Different ML models, including support vector machines (SVMs), random forests, and deep learning methods, are used to improve system performance. Wind power forecasting is a vital part of grid stability and energy planning, with ML algorithms using historical and current meteorological data to enhance accuracy. Fault detection and diagnosis by ML identifies abnormalities in wind turbines at a lower cost, preventing faults. Deep learning algorithms, including CNNs and LSTM networks, are increasingly used for high-level data-driven analysis. In addition, ML improves turbine operation by enabling real-time parameter adjustments, such as blade angle and rotor speed, and energy management systems use ML to improve power distribution efficiency. However, problems including data availability, computational difficulty, and model interpretability remain. The chapter elaborates on these points and outlines avenues for future research, serving as an all-around resource for wind energy system researchers and engineers.

Keywords: Machine learning applications, Predictive maintenance, Renewable energy optimization, Smart grid integration, Wind power forecasting.

INTRODUCTION

Wind energy is one of the world's most rapidly growing renewable energy sources, contributing significantly to reducing fossil fuel use, mitigating green-

* **Corresponding author Inam Ul Haq:** School of Engineering and Technology (SET), CGC University, Mohali, Punjab, India; E-mail: mirinam525@gmail.com

Sugandha Singh, Inam Ul Haq, Sachin Ahuja, Akib Mohi Ud Din Khanday & Neha Sharma (Eds.)

house gas emissions, and ensuring sustainability. As evident in Fig. (**1**), the World Wind Power Projection (2020–2030) indicates rising the uptake of onshore and offshore wind power, which underscores its growing role in the energy transformation [1]. Wind power is clean and abundant in contrast to conventional sources such as coal or gas. Despite this, the challenges of variable wind speeds, uncertain weather, and effective delivery of energy remain. These factors require the implementation of cutting-edge technologies, including machine learning, to achieve optimal efficiency and reliability in wind power generation.

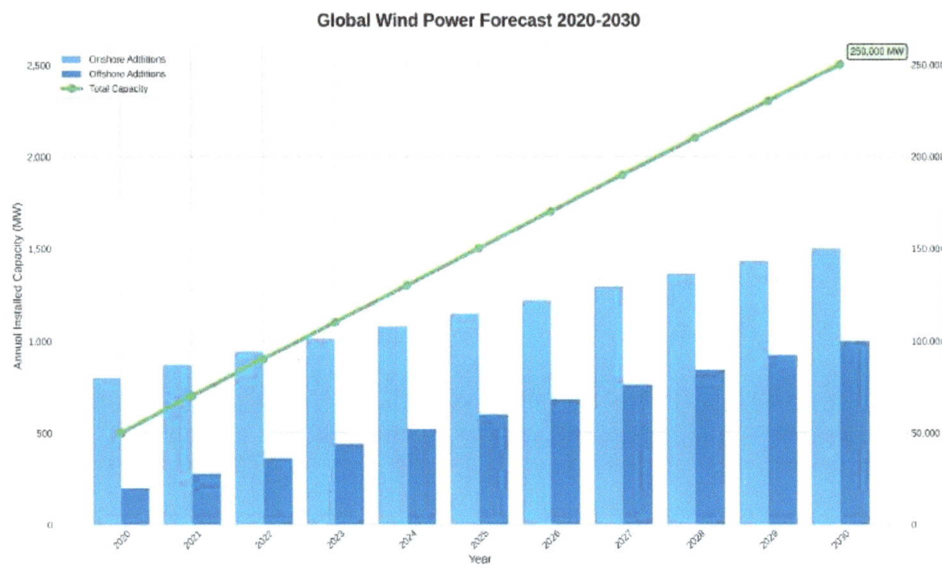

Fig. (1). "Global Wind Power Forecast (2020–2030): Projected annual installed capacity (MW) for onshore and offshore wind energy, along with cumulative total installed capacity (GW)."

Machine learning (ML) is also coming into the picture as a potential solution to address such issues. ML addresses the problem of training computer networks to learn from data and subsequently make decisions. With deep analysis of historical and real-time data, ML models can assist across all facets of wind energy, ranging from wind power generation forecasting and fault detection in turbines to operation optimization and grid integration. Wind power forecasting is one of the key uses of ML in wind power. It is not possible to predict the power output of a wind farm at any given time since the wind speed is always fluctuating. Deep learning models and neural networks can be trained with historical wind data and weather conditions to predict power output with high accuracy. Grid managers can make more effective plans, prevent energy wastage, and prevent blackouts.

Fault prediction and predictive maintenance is another important application. Wind turbines are expensive and operate in harsh conditions. If a turbine fails prematurely, it would be very expensive to repair and lead to a loss of energy output. ML algorithms can monitor the operation of a turbine in real time, detecting potential faults before they become critical. This enables servicing to be done at the right time, which is cost-effective and ensures continuous energy delivery.

In addition, ML is applied for the optimization of turbine operation. Rotor speed, blade pitch angle, and wind direction are key control variables essential for turbine efficiency. Sophisticated ML algorithms can dynamically control these variables to deliver maximum power while minimizing wear and tear on the machine.

ML is also beneficial for regulating energy distribution and integrating wind energy into the grid. As wind power is intermittent, it is possible to forecast it through ML-based intelligent grid systems; excess energy can be stored, and supply can be regulated efficiently. This enhances the stability of the power grid and increases the reliability of renewable energy sources.

This chapter will cover various ways in which ML is transforming the wind energy industry. It will outline some of the ML techniques applied for wind power forecasting, fault detection, turbine tuning, and energy control. It will also address issues such as data availability, computational costs, and the explainability of AI systems. Finally, future directions and research avenues will be discussed, providing an outlook on how ML will continue to transform wind energy systems.

MACHINE LEARNING APPLICATIONS IN WIND ENERGY SYSTEMS

Wind Power Forecasting

For dispatch optimization, grid stability, and economic optimization of wind farms, precise wind power forecasting is extremely crucial. Since wind resources are stochastic and intermittent, precise forecasts cannot be obtained through conventional statistical techniques. There has been evidence that it is extremely beneficial to employ Machine Learning (ML) algorithms with the assistance of big data in enhancing the accuracy of prediction, detecting intricate patterns, and re-calibrating for changing weather patterns [2].

Importance of Wind Power Forecasting

Wind power is highly sensitive to irregular meteorological conditions, and thus, it is not easy to predict power generation with high precision. Accurate forecasting is of great importance to:

- **Grid Stability:** Inaccurate predictions can lead to grid imbalance, resulting in frequency and voltage fluctuation.
- **Economic Planning:** Prediction allows energy producers to maximize trading strategy in energy markets, lowering operating costs.
- **Energy Storage and Dispatch:** Accurate forecasts facilitate better coordination of demand-response systems and battery storage.
- **Turbine and System Optimization:** Efficiency improvement and maintenance planning are achievable for operators with predictions.

ML ALGORITHMS FOR WIND POWER FORECASTING

ML models are capable of extracting hidden dependencies from historical data and meteorological inputs such as wind speed, direction, temperature, and air pressure [3]. Various machine learning algorithms have been utilized to predict energy output with differing levels of accuracy and efficiency. The table below provides a comparative analysis of some commonly employed ML algorithms, highlighting their advantages, limitations, and examples from recent research [4]. Table **1** provides a comparative analysis of various machine learning algorithms used in wind power forecasting, highlighting their accuracy, forecast horizons, and operational trade-offs. Table **1** provides a comparative analysis of various machine learning algorithms used in wind power forecasting, highlighting their accuracy, forecast horizons, and operational trade-offs.

Table 1. Comparative analysis of ML algorithms in wind power forecasting.

ML Algorithm	Strengths	Weaknesses	Reference
ANN(Artificial Neural Networks)	Captures non-linear patterns, flexible	Needs big training data, computationally costly.	[5]
LSTM(Long Short-Term Memory)	Handles sequential dependencies well, suitable for time-series data.	Risk of overfitting, need for hyperparameter tuning.	[6]
XG Boost	High accuracy, interpretable, efficient for structured data	Works less well with highly dynamic data.	[7]
Hybrid CNN- LSTM	Merges spatial and temporal characteristics for enhanced accuracy.	High computation, needs huge datasets	[8]

Recurrent Neural Networks (RNNs)

RNNs can also be used for time series forecasting because they effectively handle sequential data. They can recall previous states and, therefore, manage temporal dependencies in wind power data. However, standard RNNs face short-term memory loss due to vanishing gradients, which limits their ability to make long-range predictions [9].

Long Short-Term Memory (LSTM) Networks

LSTMs, which are advanced versions of RNNs, address the limitations of basic recurrent networks by incorporating memory cells that selectively remember and forget data over extended periods. LSTMs have shown strong performance in extracting long-term dependencies in time-series data, making them suitable for wind power forecasting. They can address seasonality and sudden changes in data, and their predictive capability surpasses that of basic models [10].

Gradient Boosting Algorithms

Gradient boosting models, such as XGBoost, LightGBM, and CatBoost, have gained popularity due to their ability to handle large datasets and provide precise predictions. These models work by iteratively minimizing prediction errors, combining weak learners into a strong ensemble model. Unlike deep learning models, gradient boosting models utilize relatively low computational power and offer interpretability, making them an attractive alternative for wind power forecasting [11].

Hybrid ML Models

One of the emerging trends in wind energy prediction is the ensemble of several ML models to enhance performance further. For instance:

- CNN-LSTM models integrate convolutional neural networks (CNNs) for spatial feature extraction and LSTMs for temporal pattern discovery.
- Ensemble models combine recurrent neural networks (RNNs), decision trees, and support vector machines (SVMs) for improved generalization and stability.
- Physics-based ML models merge ML methods with expert knowledge to enhance predictive accuracy during adverse weather conditions [12].

CASE STUDY: GOOGLE DEEPMIND'S AI FOR WIND POWER FORECASTING

One of the most significant ML-based wind power forecasting examples is Google's DeepMind project. Collaborating with a U.S. wind energy firm, DeepMind created a neural network that was trained on weather forecasts and past turbine performance to forecast wind power output 36 hours before actual generation. This method enhanced the value of wind power by around 20%, allowing enhanced grid integration and optimization of the energy market. The system advises the best hour-by-hour delivery commitments to the grid one day ahead, proving that ML is able to enhance predictability as well as the value of wind energy [13].

Fault Detection and Predictive Maintenance

Wind turbines are exposed to extreme environmental conditions, causing wear and tear over time. Sudden failures not only halt energy generation but also result in expensive repairs. Wouldn't it be great to foresee these problems before they become serious issues? That's where Machine Learning (ML) comes in.

The Need for Proactive Maintenance

Typical maintenance practices are usually based on routine check-ups or responding to failures as and when they arise. These are inefficient approaches [14]. With a predictive strategy, it is possible to save money, plan maintenance work at the right time, and reduce operational downtime. Using a predictive strategy, it is possible to: Look for early signs of trouble and repair the small things before they become big—schedule maintenance when low energy use is experienced to minimize downtime.

How ML Makes a Difference

Wind turbines have many sensors that monitor factors such as vibration, temperature, and the rotational speed at which they operate [15]. The ML algorithms then go through this data and look for patterns that could be significant of an issue. Now, we will mention some of the ML techniques employed:

Detection of Anomaly with Autoencoders

Autoencoders are taught to learn the normal functioning of a turbine and identify anomalies in performance. For instance, if a turbine begins to vibrate beyond normal parameters, an autoencoder can detect this as a fault, and that implies something is amiss with an element [16].

Decision Trees

They help determine faults by analyzing different sensor readings. For example, if there is a given temperature and vibration profile, the decision tree may recognize it as a sign of bearing wear.

Bayesian Networks

They are computational models of variable relationships based on probability to enable uncertainty reasoning. They evaluate the probability of various fault conditions based on deviations of sensor data.

Support Vector Machines (SVMs)

Support Vector Machines are supervised learning techniques applied for classification and regression. They are well-suited to separating normal and abnormal operating states by identifying the best boundary that separates various classes within the data.

Ensemble Learning Methods

In some cases, combining multiple models is the best approach. Ensemble methods consolidate knowledge from various algorithms to generate a more accurate and dependable prediction, which maximizes the entire fault detection process.

CASE STUDY: VESTAS WIND SYSTEMS – PREDICTIVE MAINTENANCE USING ML

Vestas, the world's leading wind turbine maker, has adopted cutting-edge predictive maintenance techniques to optimize turbine performance. Using predictive and preventive maintenance practices, Vestas has the capability to detect potential issues even before they appear and take measures to solve them, reducing downtime and maximizing yield for turbines globally. With this method, wind energy loss is minimized by enabling servicing during low wind conditions, while maintaining maximum business confidence.

Overcoming Challenges

While ML promises interesting solutions, certain challenges must be tackled:

Data Quality: Ascertaining that the gathered data is precise and exhaustive.

Model Transparency: Ensuring that decisions taken by ML models are understandable for maintenance crews.

Scalability: Using these models across multiple turbines of various specifications. Tackling these issues will increase ML's impact in predictive maintenance. Table **2** outlines the diverse applications of machine learning in the wind energy sector, ranging from power forecasting to fault detection and storage optimization.

Table 2. Machine learning applications in wind energy.

ML Application	Objective	Key Benefits	Study
Wind Power Forecasting	Forecast wind energy generation from meteorological data.	Enhances grid stability, improves energy dispatch, minimises curtailment.	[17]
Fault Detection & Predictive Maintenance	Detect turbine malfunctions prior to failure.	Lowers downtime, decreases maintenance cost, and extends turbine life.	[18]
Turbine Operation Optimization	Vary turbine parameters (blade pitch, rotor speed) dynamically.	Maximizes power output, and reduces mechanical stress.	[19]
Energy Management & Grid Integration	Match power demand and supply, and optimise energy storage.	Increases grid efficiency, and delivers power reliably.	[20]
Cybersecurity & Anomaly Detection	Detect and counter cyber attacks in wind energy systems.	Increases data security, and unauthorized access.	[21]

Optimization of Wind Turbine Operations

Wind turbines are engineering wonders that transform the kinetic energy of wind into electricity. To ensure their maximum efficiency, it is important to constantly regulate different operational parameters, including the pitch of the blades, the rotational speed of the turbine, and the devices that transform mechanical energy into electricity. Optimizing this is difficult with the constant change in wind conditions [22], yet, with the arrival of advanced algorithms such as Reinforcement Learning and Genetic Algorithms, we can optimize turbine operations to dynamically react to such changes.

REINFORCEMENT LEARNING: OPTIMIZING TURBINES TO LEARN

Reinforcement Learning (RL) helps wind turbines learn to adjust dynamically to environmental conditions by constantly optimizing operational parameters based on real-time feedback. In RL, an agent (here, the control system of the turbine) acts on its environment (the wind) and decides (tunes operational parameters) to reach a particular objective (maximize energy production). The agent is provided with feedback in the form of rewards or penalties depending on what it does, enabling it to learn and optimize with time [23].

For example, an RL-based system continuously adjusts blade pitch angles in response to changing wind speeds, maximizing energy production while minimizing mechanical stress. As time passes, the system becomes highly effective at making such adjustments, and this results in huge efficiency gains.

Genetic Algorithms: Creating the Best Plans

Genetic Algorithms (GAs) enhance the performance of turbines by means of selection, crossover, and mutation. GAs operate by generating a population of candidate solutions to a problem and repeatedly selecting, recombining, and mutating them towards better ones [24].

GAs can be used to optimize wind turbines by finding the best set of operating conditions to optimize energy output. GAs assist in finding ways that are not immediately apparent through traditional means by attempting various scenarios and iteratively refining solutions [25].

Combining RL and GAs: A Powerful Synergy

Current studies have investigated combining RL and GAs to even further improve the wind farm performance. For example, New studies proposed a Reinforcement Learning-Enhanced Genetic Algorithm (RLGA) that combined the adaptive learning properties of RL and the evolutionary optimization capabilities of GA to promote wind farm effectiveness and lower computer overhead [26].

PRACTICAL APPLICATIONS AND ADVANTAGES

There are various useful advantages to operating wind turbines with these sophisticated algorithms:

- **Increased energy production:** The best wind-harvesting capacity is provided *via* dynamic tweaking.
- **Longer Turbine Life:** By offering ideal control, this reduces mechanical wear.
- **Low-Cost Operations:** There is minimal maintenance and hence minimal downtime.

In maximizing the operation of wind turbines, the combination of genetic algorithms and reinforcement learning is a significant achievement [27]. Allowing turbines to learn from nature and incorporate the best practices is the key to tapping into wind energy and building a better future.

Grid Integration and Energy Management

Integrating wind energy into the grid is challenging due to its volatility. To have a stable and consistent supply of electricity, balancing the variable generation of wind power with the changing grid demand is necessary. That is where Machine Learning (ML) comes into play, providing new ways of managing energy more effectively [28].

Balancing Act: Supply and Demand

Consider the power grid to be a busy road, and electricity to be the traffic traveling to service homes, businesses, and industries [29]. Adding wind power is similar to adding a collection of cars whose speeds fluctuate with changing wind speeds. We must have intelligent systems that can predict and act on these fluctuations to avoid traffic congestion (or in this context, power loss).

Smart Grids: The Intuition Behind Operation

Smart grids incorporate IoT sensors, real-time analysis, and ML-based optimization to improve energy prediction, load balancing, and grid stability.

These grids can:

Forecast Energy Generation: By analyzing historical records and weather forecasts, ML models can forecast the amount of wind energy to be generated at any given time.

Control Energy Storage: Any surplus wind energy can be stored in batteries or other storage systems. ML helps determine the best times to store or dispense this energy so it is available at the right moment.

Maximize Power Dispatch: Choosing which power sources to draw from and when is essential. ML algorithms help make these choices, ensuring the most cost-efficient and effective sources are used.

DEEP REINFORCEMENT LEARNING: LEARNING THE ROPES

Among the advanced ML methods trending in energy management is Deep Reinforcement Learning (DRL). Deep Reinforcement Learning (DRL) enables systems to learn from their environment and optimize decision-making through real-time feedback and self-adjusting learning. In the field of energy management:

Home Energy Systems: DRL maximizes energy usage in intelligent homes by monitoring past usage patterns, forecasting demand variations, and automatically adjusting power allocation to reduce costs without compromising user comfort.

Grid Management: Grid management systems based on DRL dynamically balance supply and demand by forecasting variations in wind power generation, optimizing storage usage, and stabilizing grid performance.

A review identifies the application of reinforcement learning to the development of Home Energy Management Systems (HEMS) and its potential for adaptive management of real-time, complex energy requirements.

Adaptive Clustering: Grouping for Efficiency

Adaptive clustering optimizes energy management by dividing consumers and generation profiles into similar categories, facilitating enhanced demand response and resource distribution [30].

For instance:

Consumer Clustering: By grouping consumers with identical energy consumption habits, utilities are able to craft demand response programs that incentivize energy consumption off-peak hours, evening out the load on the grid [31].

Resource Matching: Clustering identifies which energy resources can be matched together to supplement one another so that there is a consistent supply of power [32].

Research shows that incorporating adaptive clustering into reinforcement learning maximizes the performance of decentralized energy systems, such as solar-powered wireless sensor networks, through real-time energy allocation optimization.

CHALLENGES AND FUTURE DIRECTIONS IN ML FOR WIND ENERGY

Challenges in Using ML in Wind Energy

Data Quality and Availability: ML models are based on high-quality, high-frequency data from wind turbines, weather sensors, and power grids. Inconsistencies can be introduced by missing entries, faulty sensors, and communication errors, thereby decreasing the precision of predictions. Improving ML model reliability is dependent on addressing these data constraints.

Computational Expenses: Deep learning models require significant computational power, leading to increased energy consumption and operational costs. This becomes an obstacle for smaller organizations due to their limited budgets, rendering cost-effective model optimization an important research topic.

Model Interpretability: The transparency of opaque ML models ('black box' problem) decreases decision-making transparency, and it becomes difficult for wind farm operators to have confidence in and utilize AI-based systems.

Scalability and Flexibility: Different regions and topographies have relatively differing wind conditions. Because models have to be tuned to local variables in order to be accurate, there is no way to train one machine learning model that will work best for multiple wind farms. Table **3** summarizes the main challenges associated with applying machine learning in wind energy systems and the corresponding mitigation strategies.

Table 3. Machine learning challenges and solutions.

Challenge	Description	Proposed Solution
Data Quality & Availability	Missing data, sensor faults, inconsistent data.	Utilize data augmentation, preprocessing.
Computational Costs	High power requirements for deep learning models.	Efficient algorithms, cloud computing.
Model Interpretability	Black-box behavior of ML models erodes confidence	Explainable AI (XAI)
Scalability	Models don't generalize to multiple wind farms	Transfer learning, federated learning.
Cybersecurity Risks	Wind energy systems are susceptible to cyberattacks.	AI-based anomaly detection.

FUTURE DIRECTIONS FOR RESEARCH IN ML FOR WIND ENERGY

- **Hybrid ML Models:** Merging physics-based models with ML methods improves wind prediction and control in turbines by incorporating knowledge of the domain with data-driven information, making predictions more reliable. This integration utilizes the advantages of both approaches, achieving more trustworthy predictions by adding physical laws and data-driven insights.
- **Federated Learning:** Federated Learning maintains data privacy and confidentiality but facilitates decentralized training of machine learning on various wind farms with collaborative model improvement. It improves data confidentiality and enables users to leverage many different datasets to develop more general models.

- **Quantum Machine Learning:**It uses quantum computing to enhance ML model training and real-time prediction of wind energy generation, achieving energy maximization beyond the computational power of conventional systems. Although still in its early stages, this application has the potential to perform more complex calculations more efficiently than conventional computers.
- **Explainable AI (XAI):** XAI enhances the use of it and promotes assurance through improving the transparency of machine learning models so that wind farm operators comprehend decision-making. In order to facilitate greater understanding and control of wind energy systems, XAI offers transparent explanations of forecasts and choices.
- **Integration of Edge AI with IoT:** Maximum efficiency, decentralization of decision-making, and real-time wind farm monitoring are ensured by the compatibility of ML, Edge AI, and IoT, as well as the removal of reliance on cloud processing. Along with greater efficiency and quicker responses to changing conditions, there is reduced dependence on centralized data processing.

CONCLUSION

With increased power generation, efficient management, and improved operation efficiency, machine learning (ML) revolutionized wind energy systems and opened the way for smart renewable energy. Better forecasting of the winds results in better warning systems and other maintenance procedures that save downtime. More electricity can be generated by turbines by doing so more efficiently due to more advanced machine learning technology. The stability and dependability of renewable energy in modern power grids have also been significantly boosted thanks to ML-based energy management and grid integration.

Despite this, challenges in data quality, computational cost, model interpretability, and security threats need to be addressed to facilitate seamless integration of machine learning into wind power generation. Nevertheless, research being conducted on hybrid ML models, federated learning, quantum computing, and explainable AI (XAI) is very promising for further advances. With these new technologies, the wind power industry can develop further, as renewable energy will become more efficient, cost-reducing, and sustainable.

In the years to come, the combination of ML, IoT, and edge computing will continue to improve real-time decision-making for wind farms. With the growing global drive towards clean energy, the use of ML in wind power will increase, leading the way to intelligent and more resilient renewable energy systems. Addressing these challenges and capitalizing on nascent ML advancements will propel the world toward a cleaner, more resilient, and smarter wind power system.

REFERENCES

[1] D. Kaushik, P. Gulia, N. S. Gill, M. Yahya, P. K. Shukla, and J. Shreyas, "Synergizing blockchain and AI to fortify IoT security: a comprehensive review", *Artif. Intell. Rev.,* vol. 59, no. 2, p. 41, 2026. [http://dx.doi.org/10.1007/s10462-025-11434-0]

[2] Y. Wang, Q. Zhang, and X. Guo, "Machine learning methods for wind power forecasting: A review", *Appl. Energy,* vol. 268, p. 115074, 2020.

[3] I. Ul Haq, "Diagnosis of Neurological Disease Using Bioinspired Algorithms", In: *Bio-Inspired Optimization for Medical Data Mining,* S. Srivastava, A. Anand, A. Kumar, B. Saini, P. S. Rathore, Eds., Wiley-IEEE Press: Hoboken, NJ, USA, 2024, pp. 245-270. [http://dx.doi.org/10.1002/9781394213894.ch11] [http://dx.doi.org/10.1002/9781394214211.ch12]

[4] B. Mohandes, M. S. El Moursi, N. Hatziargyriou, and S. El Khatib, "A review of power system flexibility with high penetration of renewables", *IEEE Trans. Power Syst.,* vol. 34, no. 4, pp. 3140-3155, 2019. [http://dx.doi.org/10.1109/TPWRS.2019.2897727]

[5] X. Deng, H. Shao, C. Hu, D. Jiang, and Y. Jiang, "Wind power forecasting methods based on deep learning: A survey", *Comput. Model. Eng. Sci.,* vol. 122, no. 1, pp. 273-302, 2020. [http://dx.doi.org/10.32604/cmes.2020.08768]

[6] H. Alkhoder and M. Al-Razgan, "An Intelligent Framework for Learnability Prediction Using Enhanced Cognitive Walkthroughs and Machine Learning," *IEEE Access*, vol. 14, pp. 6619–6634, 2026.

[7] I. U. Haq, A. Kumar, and P. S. Rathore, "Machine learning approaches for wind power forecasting: a comprehensive review", *Discov. Appl. Sci.,* vol. 7, no. 10, p. 1139, 2025. [http://dx.doi.org/10.1007/s42452-025-07675-x]

[8] M.Q. Raza, and A. Khosravi, "A review on artificial intelligence based load demand forecasting techniques for smart grid and buildings", *Renew. Sustain. Energy Rev.,* vol. 50, pp. 1352-1372, 2015. [http://dx.doi.org/10.1016/j.rser.2015.04.065]

[9] T. Kabir, "Intelligent Condition Monitoring and Fault Diagnosis of Electrical Power and Control Systems Using Machine Learning–Based Predictive Analytics," *Am. J. Interdiscip. Stud.*, vol. 7, no. 1, pp. 177–222, Jan. 2026.

[10] I. Ul Haq, G. Kaur, and A. H. Rather, "Fruit Leaf Classification Using Transfer Learning for Automation and Industrial Applications", In: *Deep Learning Techniques for Automation and Industrial Applications,* P. S. Rathore, S. Ahuja, S. R. Burri, A. Khunteta, A. Baliyan, A. Kumar, Eds., Wiley-IEEE Press: Hoboken, NJ, USA, 2024, pp. 115-138. [http://dx.doi.org/10.1002/9781394263431.ch6] [http://dx.doi.org/10.1002/9781394234271.ch10]

[11] M.V. Ramesh, "Design and development of a wireless sensor network for detecting rainfall-induced landslides", *Ad Hoc Netw.,* vol. 13, pp. 2-18, 2014. [http://dx.doi.org/10.1016/j.adhoc.2012.09.002]

[12] I. U. Haq, and A. M. U. D. Khanday, "Exploring the Role of Explainable AI in Women's Health: Challenges and Solutions", In: *Explainable Artificial Intelligence in the Healthcare Industry,* A. Kumar, T. A. Kumar, P. Das, C. Sharma, A. K. Dubey, Eds., Wiley-IEEE Press: Hoboken, NJ, USA, 2025, pp. 283-306. [http://dx.doi.org/10.1002/9781394249312.ch13]

[13] Z. Wang, W. Xue, and J. Zhang, "Hybrid optimization of wind turbine parameters using reinforcement learning and genetic algorithms", *Energy Convers. Manage.,* vol. 237, p. 114032, 2021.

[14] I. Ul Haq, I. U. Hassan, and H. A. Shah, "Machine Learning Techniques for Result Prediction of One Day International (ODI) Cricket Match", In: *Proceedings of the 2023 IEEE 8th International Conference on Convergence Systems and Innovative Technologies (I2CT)*, 2023. [http://dx.doi.org/10.1109/I2CT57861.2023.10126241]

[15] P.J. Werbos, "Backpropagation through time: what it does and how to do it", In: *Proceedings of the IEEE,* vol. 78, no. 10, pp. 1550-1560, 1990. [http://dx.doi.org/10.1109/5.58337]

[16] J. Liu, J. Liu, K. Lin, D. Suo, B. Zhang, and S. Liu, "A three-stage transfer learning framework for transient stability assessment of power systems," *Electr. Power Syst. Res.*, vol. 253, p. 112471, Jan. 2026.

[17] B. Bhardwaj, and M. Ganesan, "Machine Learning in Wind Energy: Generation to Supply", In: *Machine Learning Techniques for Smart City Applications: Trends and Solutions.,* D.J. Hemanth, Ed., Springer: Cham, 2022, pp. 215-229. [http://dx.doi.org/10.1007/978-3-031-08859-9_12]

[18] W. Hu, G. Zhang, Z. Zhang, S. Abulanwar, and F. Blaabjerg, "AI for Power Electronics and Renewable Energy Systems", In: *Wiley-IEEE Press* 1st. Hoboken, NJ, USA: Wiley-IEEE Press, 2024.

[19] "Wind Energy Storage and Conversion: From Basics to Utilities", In: *Scrivener Publishing LLC*Beverly, MA, USA, 2024.

[20] W. Fan, Y. Fu, S. Sheng, J. Bian, Y. Zhou, and H. Xiong, "DEWP: Deep Expansion Learning for Wind Power Forecasting", *ACM Trans. Knowl. Discov. Data,* vol. 18, no. 6, pp. 1-21, 2024. [http://dx.doi.org/10.1145/3637552]

[21] J. Chatterjee, and N. Dethlefs, "Scientometric review of artificial intelligence for operations & maintenance of wind turbines: The past, present and future", *Renew. Sustain. Energy Rev.*. vol. 144, Art. no. 111051, July 2021. [http://dx.doi.org/10.1016/j.rser.2021.111051]

[22] C. Moss, R. Maulik, and G. V. Iungo, "Modeling Wind Turbine Performance and Wake Interactions with Machine Learning", *arXiv preprint arXiv:2212.01483,* 2022.

[23] H. Peng, X. Xie, and T. Zou, "Neural Network-Based wind speed Prediction and Evaluation Using the NREL Wind Integration National Dataset," *Informatica*, vol. 50, no. 6, pp. 433–454, 2026.

[24] J. Liu, Y. Liu, Q. Shen, K. Li, and Y. Chen, "A hybrid artificial intelligence and deep learning architecture for accurate renewable energy forecasting: comprehensive case studies on wind and PV power," *Sustain. Energy Technol. Assess.*, vol. 85, p. 104812, Jan. 2026.

[25] A. Eljyidi, H. Jebari, S. Rekiek, and K. Reklaoui, "A Hybrid Metaheuristic and Deep Learning Framework for Robust Predictive Maintenance in Wind Energy Systems," *J. Eur. Syst. Autom.*, vol. 59, no. 1, pp. 207–217, Jan. 2026.

[26] J. Sanchez-Espiga, A. Diez-Ibarbia, F. Martin-Manrique, M. Iglesias, L. Bonaiti, A. Fernandez-de--Rincon, and F. Viadero, "Supervised Learning Neural Networks applied to the prediction of the Mesh Load Factor in a Wind Turbine Planetary Gearbox," *SSRN*, Art. no. 6544715, Jan. 2026.

[27] S. Kumar, V. K. Sharma, and D. K. Patel, "Artificial Intelligence-Driven Flow Optimization in Renewable Energy Systems," In: *Artificial Intelligence and Computational Modeling in Heat Transfer and Fluid Dynamics*, 1st ed. Hoboken, NJ: Wiley-IEEE Press, 2026, pp. 237–276.

[28] B. Wu and X. Wang, "Low-Cost, Zero-Carbon Microgrid Electricity Price Forecasting and Dispatch Strategy with Hybrid Hydrogen Gas Turbines," *Expert Syst. Appl.*, vol. 309, p. 131188, Jan. 2026.

[29] N. W. Hesty, D. A. Renata, B. Pranoto, P. T. Wijaya, R. P. Wijayanto, D. Rostyono, and A. H. Kuncoro, "A Comparative Study of Machine Learning and Kriging: Improving Wind Resource Assessment In: *Data-Scarce, Monsoon-Affected Regions*," Remote Sens. Appl. Soc. Environ., vol. 42, p. 101909, 2026.

[30] D. S. Sehat and S. S. Sehat, "Hybrid Fuzzy PID with Soft Actor-Critic Reinforcement Learning for Wind Turbine Pitch Control," *Authorea Preprints*, Jan. 2026.

[31] S. R. Jena, S. Raj, S. Agarwal, and D. K. Ram, "AI in renewable energy resources: Current trends, challenges and future prospects," In: *Artificial Intelligence and Sustainable Innovation*, 2026, pp.

606–613.

[32] G. S. Negi, H. Mohan, M. K. Gupta, R. Singh, A. Gehlot, A. K. Thakur, S. Dogra, and L. R. Gupta, "Leveraging machine learning for optimized microgrid management: Advances, applications, challenges, and future directions," *Renew. Sustain. Energy Rev.*, vol. 226, p. 116345, Jan. 2026.

CHAPTER 7

AI for Wind Turbine Performance and Maintenance

Harsh Kumar[1] and **Malik Muzamil Ishaq**[1,*]

[1] *Department of Computer Science & Engineering, Akal University, Talwandi Sabo, Punjab, India*

Abstract: A fundamental aspect of switching to renewable energy on a global scale is wind energy. However, due to high operating costs, wear, and unpredictable weather, the maintenance and performance optimization of wind turbines remain critical problems. Conventional maintenance practices are based on reactive repair or planned inspections, which lead to unplanned downtime and lost revenue. Artificial Intelligence (AI) has empowered the optimization and maintenance of wind turbine performance. AI ensures real-time monitoring and fault detection through the use of machine learning algorithms, deep learning, and Internet of Things-based predictive analytics. Predictive maintenance, supported by AI models that analyze sensor data, is able to promote early fault detection, lower maintenance costs, and extend turbine life. In addition, costly manual blade inspections are not necessary, as the use of AI-powered drones equipped with computer vision can take care of these tasks. In this chapter, the application of AI to predictive maintenance and wind turbine performance monitoring is discussed. It discusses a number of AI-driven techniques, including energy output prediction, vibration analysis, and anomaly detection. Real-world case studies are used to show how AI is transforming wind energy maintenance. The chapter also lists the primary barriers to the implementation of AI, including data availability, computing expenses, and integration with existing infrastructure at wind farms.

Keywords: Artificial Intelligence (AI), Digital twin technology, Energy forecasting, Energy grid integration, Fault detection, Hybrid renewable energy systems, Predictive maintenance, Wind energy, Wind turbine optimization.

INTRODUCTION

Overview of Wind Energy and Its Growing Importance

The renewable energy source that is growing the fastest in the world today is wind power. It is an important element of the international efforts aimed at reducing climate change and dependence on fossil fuels [1]. Nearly all countries in the

* **Corresponding author Malik Muzamil Ishaq:** Department of Computer Science & Engineering, Akal University, Talwandi Sabo, Punjab, India; E-mail: muzamilmalik754@gmail.com

Sugandha Singh, Inam Ul Haq, Sachin Ahuja, Akib Mohi Ud Din Khanday & Neha Sharma (Eds.)

entire world are investing heavily in wind power as a cleaner, more sustainable alternative to traditional energy sources, as environmental degradation becomes a more pressing issue. The capacity of wind power is growing quickly, and many hundreds of onshore and offshore wind farms are already in operation globally, as reported by the Global Wind Energy Council. Wind turbines are necessary to catch wind energy and convert it into electricity.

These imposing buildings that have big blades work through the generation of wind energy that turns a generator producing electricity. Many factors, such as wind speed, the design of the turbines, location; and climatic conditions, varying with the locations in question, have an influence on how efficient wind turbines will be. Even though wind energy is abundant and cost-effective in the long term, operation of the turbines requires optimal efficiency that cannot be realized due to inefficiency of operations, weather changes, and mechanical wear and tear. Operators must continuously assess turbine condition and adjust energy output in order to maximize efficiency, preserve dependability, and minimize downtime [2]. Historically, reactive repairs, manual inspections, and scheduled maintenance have been used to do this. But because of technological developments, especially Artificial Intelligence (AI), new approaches are appearing that can maximize energy output, anticipate turbine breakdowns before they happen, and enhance system efficiency as a whole. The management of wind farms is being completely transformed by AI-based predictive maintenance, real-time monitoring, and intelligent control systems.

THE ROLE OF AI IN WIND ENERGY OPTIMIZATION

Performance, efficiency, and maintenance have all improved dramatically because of the integration of wind energy systems by AI. Large quantities of data obtained from the wind turbines can be analyzed in real time by AI algorithms, thus supporting accurate forecasting, better decision-making [3], and better operations.

Some of the key ways AI is contributing to wind energy optimization include:

i. **Predictive Maintenance:** To find early indications of wear and tear, AI-based models examine sensor data from wind turbines. Maintenance staff can minimize downtime and maintenance expenses by spotting any faults early on and making repairs on schedule.
ii. **Energy Forecasting:** AI outperforms conventional forecasting algorithms in predicting wind speeds and power generation. This lessens swings and guarantees a steady supply of electricity by assisting grid operators in more efficiently integrating wind energy into the power grid.

iii. **Autonomous Control Systems:** AI maximizes energy production and efficiency by allowing wind turbines to autonomously modify their operating parameters based on current weather conditions.
iv. **Fault Detection and Diagnostics:** Before they result in significant failures, AI-powered fault detection systems can find mechanical problems in turbine parts, including gearboxes, generators, and rotor blades.
v. **Performance Optimization:** AI algorithms endeavour to bring maximum output of energy in diverse wind conditions by considering environmental inferences and past data to enhance turbine performance.

Wind farm operators can switch from reactive maintenance techniques to proactive and predictive ones thanks to AI's rapid and accurate analysis of massive datasets. This lowers expenses, prolongs the life of turbine parts, and boosts overall effectiveness. As depicted in Fig. (**1**), AI processes vibration sensor data to detect anomalies and optimize maintenance in wind turbines.

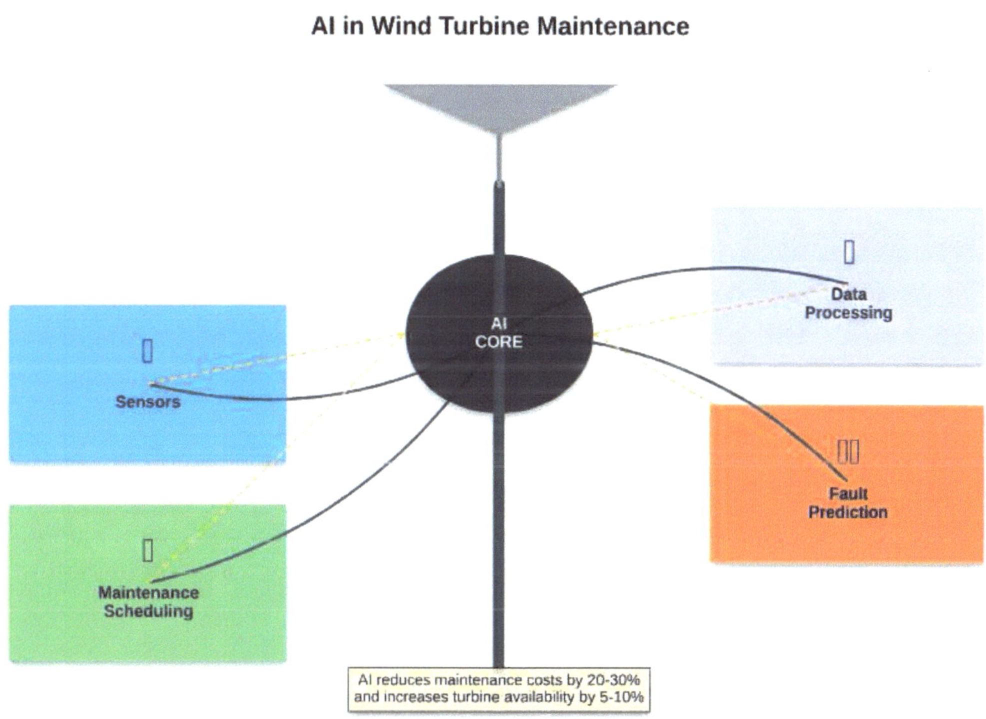

Fig. (1). Shows how integration of the AI plays out in the vibrations of wind turbines, thus requiring the use of the sensor data for real-time monitoring operations, fault predictability, as well as in optimized maintenance scheduling.

Table **1** presents a comprehensive overview of how various AI techniques are applied to optimize different aspects of wind energy systems.

Table 1. Key AI applications in wind energy optimization.

AI Application	Function	Impact on Wind Energy
Predictive Maintenance	Analyzes turbine sensor data to detect early signs of failure.	Reduces repair costs, extends turbine lifespan.
Energy Forecasting	Uses AI models to predict wind speeds and energy output.	Improves grid stability, prevents energy wastage.
Digital Twin Technology	Creates a real-time virtual replica of wind turbines.	Enhances remote monitoring and efficiency optimization.
Autonomous Control Systems	AI adjusts turbine settings in real time based on weather conditions.	Maximizes power output, reduces operational inefficiencies.
AI-Driven Blade Design	Uses AI to optimize aerodynamics and materials.	Improves energy capture, enhances turbine durability.

CHALLENGES IN WIND TURBINE OPERATIONS

Despite advancements in wind energy technology, several challenges persist in wind turbine operations:

a. **Mechanical Wear and Tear:** Because of mechanical movement, temperature changes, and wind forces, wind turbine components are constantly under stress. This causes component deterioration over time, necessitating regular maintenance.
b. **Weather Variability:** The weather has a significant impact on wind energy generation. Unpredictable energy output can result from abrupt fluctuations in wind speed, temperature, and humidity, which can affect turbine efficiency.
c. **Unscheduled Downtime:** Unexpected turbine component failures might result in lower energy output and higher operating expenses. These failures might go unnoticed until serious harm has been done if there are inadequate monitoring measures in place.
d. **High Maintenance Costs:** Conventional maintenance methods depend on reactive repairs and planned inspections, which can be costly and ineffective. Due to urgent repairs and replacement parts, unexpected failures further raise maintenance expenses.
e. **Grid Integration Issues:** It is hard to incorporate a wind energy supply into the electrical system because it is an intermittent source of power. Vacillation of wind power generation has the potential to perturb the system when there are no fine-grained forecasts and immediate adjustments.

AI-powered solutions provide a means of successfully tackling these issues. Turbine performance [4] trends can be analysed by machine learning models, which can also identify irregularities and anticipate breakdowns before they happen. AI can also optimize energy production according to current weather conditions, increasing efficiency and lowering the difficulties associated with grid interconnection.

TRADITIONAL *VS.* AI-BASED WIND TURBINE MAINTENANCE

Wind turbine maintenance typically relies on three primary strategies, each with its own strengths and limitations. The first is Reactive Maintenance, a method where components are only repaired or replaced after they fail. While this approach may seem cost-effective initially, it often leads to unexpected breakdowns, emergency repair needs, and prolonged downtime, resulting in increased operational costs and reduced energy output. The second strategy is Preventive maintenance, which involves carrying out routine maintenance tasks at predetermined intervals, regardless of the actual condition of the equipment. Although this method helps reduce the risk of sudden failures, it can lead to unnecessary labor, downtime, and replacement of parts that may still be functioning well, thereby increasing maintenance expenses without proportionate benefits. The most advanced and efficient approach is Predictive Maintenance, particularly when powered by Artificial Intelligence. This technique leverages sensors and real-time data analytics to monitor the health and performance of turbine components continuously. By identifying wear patterns, anomalies, or early signs of potential failure, AI-driven predictive systems allow maintenance teams to intervene proactively, minimizing the risk of breakdowns and optimizing the use of resources. This intelligent, data-based strategy significantly enhances reliability, reduces downtime, and lowers overall maintenance costs in modern wind energy systems.

In the world of wind energy, moving from reactive to proactive maintenance has revolutionized operations. Predictive maintenance extends the life of turbine components, maximizes resource use, and minimizes unscheduled downtimes significantly [5]. As illustrated in Fig. (**2**), AI-driven predictive maintenance reduces overall costs compared to traditional reactive strategies by anticipating failures before they escalate.

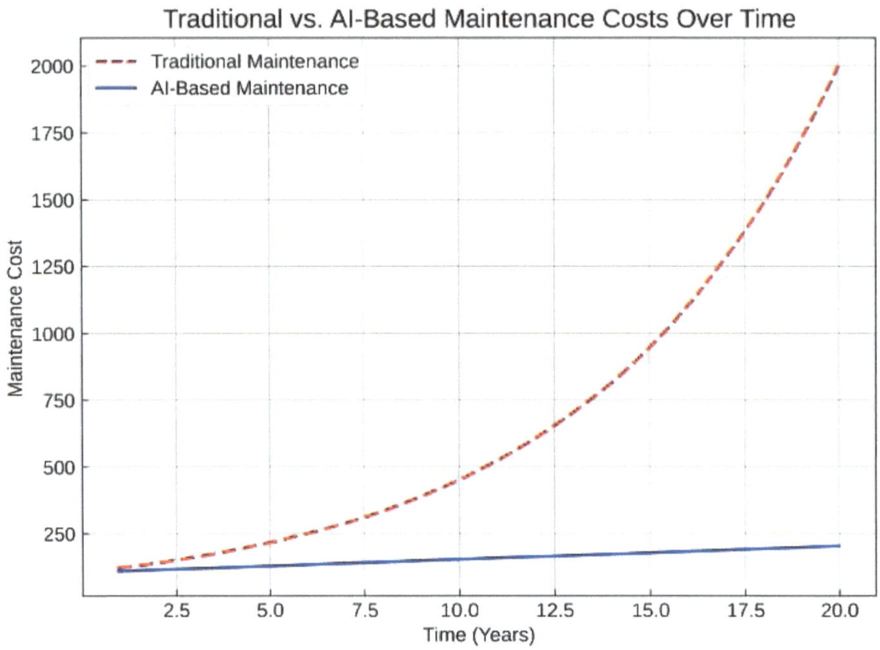

Fig. (2). Compares the cost trends of traditional reactive maintenance and AI-driven predictive maintenance. AI-based approaches result in lower long-term costs by preventing failures before they occur.

Table **2** highlights the differences between traditional and AI-based maintenance strategies, demonstrating how AI can improve efficiency and reduce costs.

Table 2. Comparison of traditional *vs.* AI-based wind turbine maintenance.

Maintenance Type	Description	Advantages	Disadvantages
Reactive Maintenance	Repairing components only after failure occurs.	Simple, minimal upfront cost.	Leads to unexpected downtime and high emergency repair costs
Preventive Maintenance	Scheduled maintenance at regular intervals regardless of component health.	Reduces unexpected failures, ensures system reliability	Can lead to unnecessary maintenance, increasing costs
Predictive Maintenance (AI-Based)	AI monitors turbine data to predict failures before they happen.	Minimizes downtime, optimizes resources, and extends turbine lifespan.	Requires investment in AI models, sensors, and cloud infrastructure.

Fig. (**3**) compares maintenance costs over time, highlighting the cost savings achieved through AI-based predictive maintenance relative to traditional approaches.

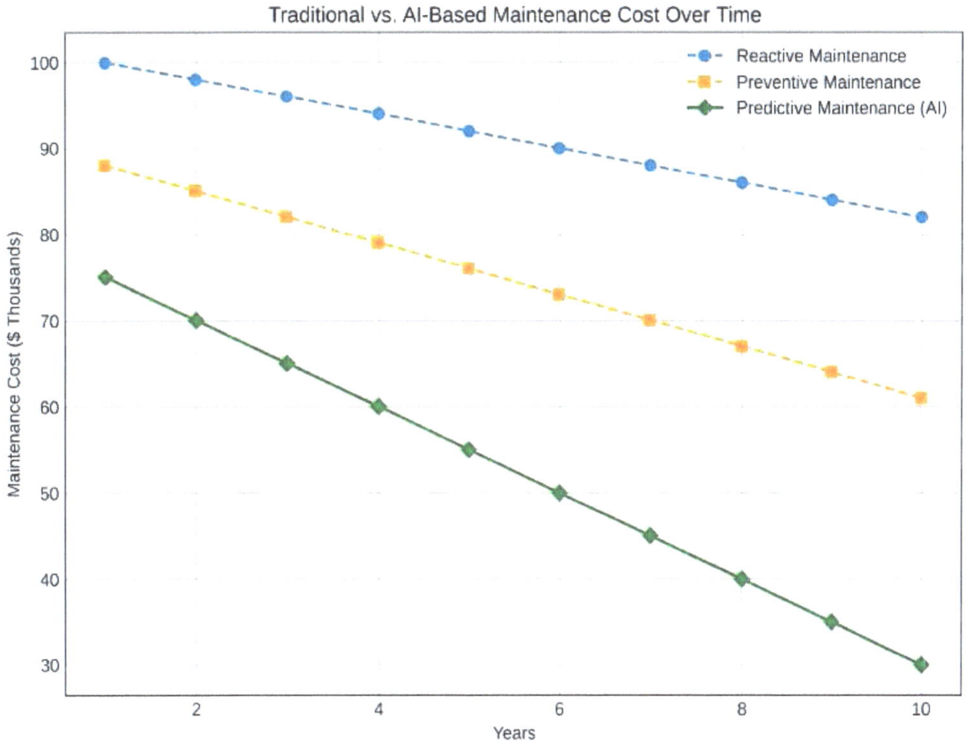

Fig. (3). Traditional *vs.* AI-based maintenance cost over time.

THE FUTURE OF AI IN WIND TURBINE PERFORMANCE OPTIMIZATION

The development of AI in optimizing wind turbines seems very promising, with continuous improvement in machine learning, IoT, and big data analytics. Some of the important trends projected to affect the future of AI in wind energy include:

i. **Integration of Digital Twins:** Virtual representations of actual wind turbines that mimic their performance in the real world are called digital twins. Operators may test different scenarios, improve performance, and more precisely predict errors with AI-powered digital twins.

ii. **Enhanced AI Algorithms for Energy Forecasting:** Energy forecasting will advance with the sophistication of AI models, enabling improved energy management and grid integration.

iii. **Autonomous Maintenance Drones:** Turbine inspections will be automated using AI-powered drones fitted with cameras and sensors, eliminating the need for manual inspections and increasing maintenance effectiveness.

iv. **Self-Healing Wind Turbines:** Future AI systems might make it possible for turbines to self-correct and self-adjust small problems, which would further lower maintenance costs and increase turbine lifespan.
v. **AI-Enabled Smart Grids:** AI development in the optimization of wind turbines looks quite promising at the moment, with continued incremental improvements in machine learning, IoT, and big data analytics.

AI integration in wind turbine performance optimization will keep increasing the sustainability, efficiency, and cost-effectiveness.

One of the critical parts in the world's transition to renewable energy is wind energy. Mechanical wear, state of weather, and high operating cost, therefore, make it difficult to ensure optimal performance and maintenance of wind turbines [6]. New ways of optimizing the efficiency of wind turbines are presented by AI-driven solutions, such as intelligent control systems, real-time monitoring, and predictive maintenance.

Energy production can be enhanced by AI, maintenance costs can be reduced, and downtime can be minimized by wind farm operators through the use of AI. Digital twins, among other techniques and self-healing turbines, are just several of the more sophisticated tools the wind energy industry will acquire as AI technology has developed further. A smooth integration of AI should be the key to the future optimization of wind energy, proving a more economical, efficient, and sustainable renewable energy ecosystem.

LITERATURE REVIEW

Wind energy has become an indispensable aspect of the world's renewable energy approaches, requiring optimal performance monitoring and maintenance of wind turbines. Traditional maintenance strategy, including diesel generation, reactive and preventive methods, all have high operational cost and downtime [7]. Artificial Intelligence (AI) has risen as a strong tool in alleviating these challenges by facilitating predictive maintenance, fault detection, and performance optimisation. The recent achievements in AI-based methods for wind turbine monitoring, failure prediction, and efficiency improvements are the focus of this literature review. Predictive maintenance (PdM) uses Artificial Intelligence and Machine Learning models to use real-time data within the system, detect anomalies, predict failures before they occur, and thus reduce downtime and lower maintenance costs [8].

Machine Learning for Predictive Maintenance

It involves the application of various techniques to enhance the reliability and efficiency of wind turbines. In the realm of supervised learning, algorithms like Random Forest, Support Vector Machines, and Neural Networks are commonly used to classify the health condition of turbine components and predict potential failures. On the other hand, unsupervised learning methods such as k-means clustering and Principal Component Analysis (PCA) help detect anomalies in operational data, offering early warnings for irregularities. Deep learning approaches, particularly Recurrent Neural Networks (RNN) and Long Short-Term Memory (LSTM) models, are employed to analyze time-series data, enabling the identification of failure patterns and long-term trends. These machine learning strategies collectively contribute to proactive and data-driven maintenance planning. AI-driven condition monitoring leverages various sensors, including:

Vibration Sensors: Detect irregularities in mechanical components such as gearboxes and bearings.

Temperature Sensors: Identify overheating issues in electrical systems.

Acoustic Sensors: Capture abnormal noise patterns indicative of structural issues.

AI for Fault Detection and Diagnostics

AI-based fault detection methods enable early identification and mitigation of operational issues.

- **Anomaly Detection Techniques**

Advanced anomaly detection methods are crucial for identifying irregularities in wind turbine operations. One such method is the use of Autoencoders, which are deep learning models trained to replicate standard operational data and flag any significant deviations from this norm, thereby indicating potential faults. Another effective approach is the Isolation Forest, which works by isolating anomalies based on how easily data points can be separated from the rest of the dataset, making it efficient in detecting outliers. Additionally, Gaussian Mixture Models (GMM) serve as probabilistic models that analyze the distribution of operational states and classify them as either normal or faulty, providing valuable insights for early fault detection and maintenance planning.

- **Digital Twin Technology**

Digital twins serve as virtual models of actual wind turbines and are continuously synchronized with real-time data from their physical counterparts. These AI-powered systems allow for the simulation of turbine performance under diverse operating conditions, enabling engineers to assess and prepare for potential scenarios. Additionally, they support early failure prediction and help schedule maintenance activities more efficiently, reducing unexpected downtimes. By incorporating real-time data, digital twins also enhance the overall operational efficiency of wind turbines through continuous monitoring and adaptive adjustments.

AI FOR WIND TURBINE PERFORMANCE OPTIMIZATION

AI optimizes wind turbine performance by dynamically adjusting operational parameters.

- **Energy Forecasting**

Accurate energy forecasting is essential for the effective integration of wind energy into the power grid. Artificial Intelligence models like Long Short-Term Memory (LSTM) networks and Gated Recurrent Units (GRU) significantly enhance the prediction of wind speed and power output by analysing complex time-series data. In addition, hybrid AI models that merge physics-based simulations with machine learning techniques offer even greater forecast precision by leveraging both theoretical understanding and data-driven insights, thereby supporting more reliable and efficient energy planning [10].

- **Autonomous Control Systems**

AI-driven control systems play a crucial role in enhancing the efficiency of wind turbines by enabling intelligent, real-time adjustments. These systems automatically fine-tune blade pitch angles based on changing wind conditions, which helps maintain optimal aerodynamic performance. By minimizing unnecessary mechanical strain, they contribute to extending the operational lifespan of turbine components. Additionally, the integration of real-time feedback loops allows for adaptive energy optimization, ensuring that turbines operate at peak efficiency while responding dynamically to environmental variations.

Challenges and Future Trends

Despite its benefits, AI adoption in wind turbine maintenance and performance optimization faces several challenges [10]. As shown in Fig. (4), AI techniques enable early fault detection in essential turbine parts, facilitating timely maintenance actions.

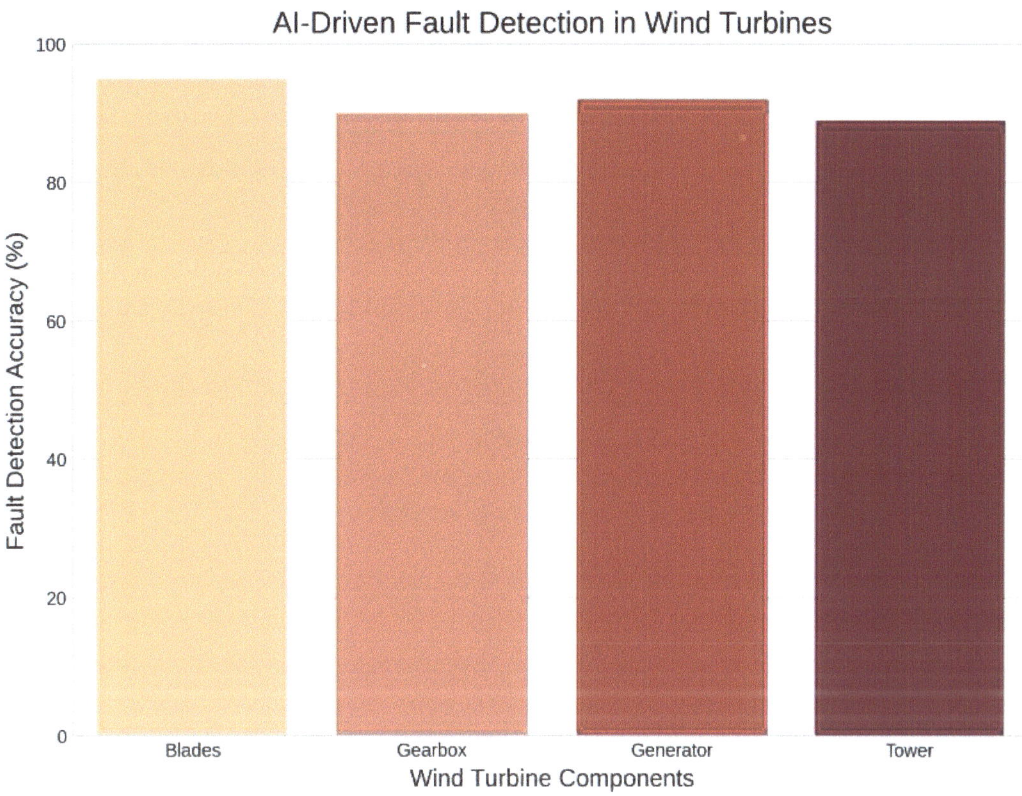

Fig. (4). Demonstrates AI's capability to detect faults in critical turbine components such as blades, gearboxes, and generators, improving maintenance efficiency and reducing downtime.

AI has revolutionized wind turbine performance monitoring and maintenance through predictive analytics, real-time fault detection, and autonomous optimization. Continued advancements in AI will drive further efficiency improvements, cost reductions, and sustainability in the wind energy sector. Future research should focus on addressing existing challenges and exploring novel AI applications for enhancing wind turbine reliability and performance.

PREDICTIVE MAINTENANCE FOR WIND TURBINES

A revolutionary technique known as predictive maintenance (PdM) applies sophisticated analytics as well as machine learning (ML) and Artificial Intelligence (AI) to monitor the health of wind turbines and detect the impending failures that will occur (before they occur). Predictive maintenance attempts to minimize downtime, save operating costs, and optimize energy output compared to maintenance strategies such as reactive maintenance, which repairs components after they fail. In preventive maintenance, components are inspected on a regular basis irrespective of the component's condition [11].

Wind turbines have to be able to meet harsh environmental conditions, including wind, temperature changes, and mechanical load. Such parts as bearings, generators, gearboxes, and blades degrade with time and may indeed fail. Anticipation of failure precursors allows predictive maintenance to prevent malfunction before it gets to the destructive stage and hence impede precise and timely interventions.

Advancements in real-time data processing, IoT Sensors, and AI-powered problem Detection are influencing the growing usage of predictive maintenance in the wind energy industry. Wind farm owners may produce more power, enhance turbine dependability, and ensure a sustainable energy future by using predictive maintenance techniques [12].

KEY COMPONENTS OF PREDICTIVE MAINTENANCE

Implementing predictive maintenance in wind turbines is a multifaceted process that incorporates advanced technologies to ensure reliability and efficiency. One of the foundational components is **sensor technology**, where turbines are outfitted with devices such as vibration sensors to detect irregularities in rotating parts like gearboxes and bearings, temperature sensors to monitor overheating in electrical components, acoustic sensors to identify abnormal noise patterns signalling mechanical issues, anemometers to measure wind speed for evaluating environmental impact, and torque sensors to detect mechanical stress that could lead to structural failures. Complementing the sensor systems is **data collection and IoT integration**, where Internet of Things (IoT) devices transmit real-time sensor data to centralized monitoring systems. Cloud-based platforms then store and analyze vast volumes of operational data, while historical performance records aid in pattern recognition and predictive diagnostics.

Machine learning and AI algorithms play a critical role by analysing incoming sensor data to detect anomalies, utilizing deep learning models that continuously refine their accuracy in identifying faults. These predictive models provide early warnings for potential failures, significantly reducing the risk of unplanned downtimes. **Remote monitoring systems** enhance operational oversight, allowing operators to track turbine health from centralized control centers. These systems generate automated alerts to inform maintenance teams of issues and display key performance metrics through digital dashboards, enabling real-time, data-driven decision-making.

Lastly, **automated maintenance planning** revolutionizes how maintenance is scheduled. Instead of adhering to fixed schedules, AI-driven systems analyse component health to determine the optimal timing for intervention. This approach allows maintenance teams to prioritize turbines that require immediate attention, avoiding unnecessary checks and thereby cutting down on operational costs. Overall, this integrated framework enhances turbine performance, ensures operational continuity, and reduces long-term maintenance expenses.

- **How AI Enhances Predictive Maintenance**

AI-powered predictive maintenance offers numerous advantages compared to traditional methods. By utilizing advanced AI models, it can identify slight changes in turbine performance well before failures happen, allowing maintenance teams to take proactive measures. This approach significantly reduces unplanned downtime and lowers operational costs by preventing expensive repairs and minimizing lost energy production. Additionally, early detection helps extend the lifespan of turbine components by avoiding severe damage, while continuous AI-based diagnostics maintain efficient operation over longer periods. Safety is also enhanced as remote AI monitoring limits the need for maintenance staff to conduct risky physical inspections, enabling them to focus only on areas flagged for attention. Furthermore, maintaining turbines in optimal condition ensures they operate at peak efficiency, maximizing energy output and balancing mechanical loads to reduce wear on individual parts. Overall, AI-driven predictive maintenance leads to more reliable, cost-effective, and safer turbine operations.

- **Case Studies on Predictive Maintenance in Wind Energy**

Several real-world applications of AI-driven predictive maintenance in the wind energy sector have shown remarkable outcomes. General Electric (GE) Renewable Energy, for instance, has developed an AI-based monitoring system for wind turbines that employs advanced analytics to forecast component failures, leading to a 20% reduction in unplanned downtime and improved turbine efficiency. Similarly, Siemens Gamesa Renewable Energy has adopted machine learning algorithms for real-time fault detection, which has successfully lowered operational costs and enhanced turbine reliability. Vestas Wind Systems have integrated IoT sensors with AI-powered platforms to monitor and optimize turbine performance globally. Their system processes vast amounts of data from thousands of turbines, delivering precise failure predictions. These case studies collectively underscore the powerful impact of AI in transforming wind energy operations by boosting reliability, maximizing efficiency, and promoting sustainability.

- **Challenges in Implementing Predictive Maintenance**

While predictive maintenance in wind turbines offers notable advantages, it also comes with several challenges that need to be addressed for successful implementation. One key issue is data quality and integration, as inconsistent or incomplete sensor data can compromise the accuracy of AI models. Moreover, integrating these systems with existing SCADA frameworks demands considerable technical expertise. The initial investment is another hurdle, with the deployment of IoT sensors, cloud infrastructure, and advanced analytics requiring substantial capital, forcing wind farm operators to weigh upfront costs against long-term gains. A skilled workforce is also essential; expertise in data science, machine learning, and wind energy operations is required, and personnel must be trained accordingly. Additionally, the increased connectivity of turbines introduces cybersecurity risks, making robust data protection measures a necessity. Furthermore, external factors such as weather variability pose a challenge, as AI models must adapt to changing conditions to maintain forecasting accuracy. Nevertheless, continuous improvements in AI and cloud technologies are gradually overcoming these barriers, making predictive maintenance a more viable and cost-effective solution for the wind energy sector.

- **The Future of Predictive Maintenance in Wind Energy**

As AI and IoT technologies continue to evolve, predictive maintenance in wind energy will become even more sophisticated. Some emerging trends include:

- **Digital Twins:** Virtual replicas of wind turbines will allow for advanced simulations and failure predictions.
- **Edge Computing:** Processing sensor data at the turbine level will enable faster and more accurate real-time diagnostics.
- **Automated Drone Inspections:** AI-powered drones equipped with thermal imaging and high-resolution cameras will perform remote inspections.
- **Self-Healing Turbines:** AI-driven control systems will enable turbines to self-adjust or repair minor faults autonomously.

- **Blockchain for Secure Data Management:** Distributed ledger technology will enhance the security and transparency of predictive maintenance data.

Predictive maintenance will remain essential to optimizing wind energy operations with these developments, guaranteeing increased sustainability, lower costs, and more efficiency [13].

Because it allows for early failure detection, minimizes downtime, and maximizes energy output, predictive maintenance has revolutionized wind turbine operations. Real-time analysis of massive volumes of sensor data by AI-powered systems yields actionable insights that improve turbine performance and dependability [14].

While problems such as data integration, cybersecurity threats, and high implementation costs persist, continued breakthroughs in AI and IoT are making predictive maintenance more effective and accessible. Predictive maintenance will be a major factor in the efficiency, sustainability, and cost savings of the renewable energy industry as wind energy continues to grow on a global scale.

ROLE OF AI IN WIND TURBINE OPTIMIZATION

AI is playing a transformative role in energy forecasting, particularly in the integration of wind power into energy systems. Traditional forecasting models often fall short in capturing rapid and unpredictable weather changes, which can lead to inefficiencies in power generation and grid management. AI addresses these limitations through advanced techniques that enhance prediction accuracy and reliability. For instance, hybrid AI models that combine machine learning methods like Random Forest with deep learning architectures such as LSTM offer improved forecasting of wind energy output by capturing both short-term patterns and long-term dependencies [15]. Additionally, Physics-Informed Machine Learning (PIML) merges physical laws with data-driven models, resulting in more accurate and realistic predictions. Furthermore, AI-based uncertainty quantification allows systems to not only forecast wind speeds but also assess the degree of uncertainty in those predictions. This capability enables grid operators to make better-informed decisions, optimize energy distribution, and maintain grid stability despite the variable nature of wind energy.

Compared to conventional forecasting techniques, AI-powered forecasting models anticipate wind speeds, energy output, and fluctuations more accurately by utilizing machine learning, deep learning, and sophisticated data analytics. By using these techniques, wind farm operators may maximize energy output, make well-informed decisions, and enhance grid stability in general.

- **Traditional *vs.* AI-Based Forecasting**
  - **Traditional Methods:**
    - Rely on Numerical Weather Prediction (NWP) models, statistical regression, and time-series analysis.
    - Limited accuracy due to an inability to capture real-time wind variations and sudden weather changes.
    - Require manual parameter tuning and extensive computational resources for processing large datasets [16].
  - **AI-Based Methods:**
    - Utilize machine learning algorithms like Random Forest, Support Vector Machines (SVM), XGBoost, deep learning models like Recurrent Neural Networks (RNNs), and Long Short-Term Memory (LSTM).
    - Continuously learn from historical wind data, meteorological information, and real-time turbine performance to refine predictions.
    - Provide short-term, mid-term, and long-term wind power predictions with higher precision, improving energy planning and utilization. The enhanced forecasting accuracy demonstrated in Fig. (**5**) underscores the importance of AI techniques in managing the inherent variability of wind energy generation.

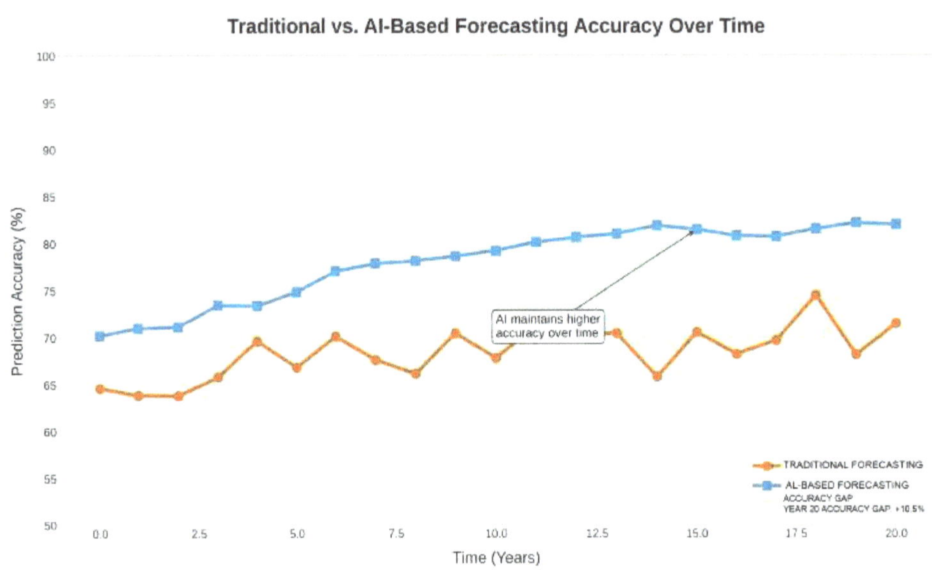

Fig. (5). Highlights the superior accuracy of AI-based forecasting models compared to traditional methods. AI enables more reliable wind power predictions, improving grid stability and energy management.

Fig. (**6**) demonstrates how AI-driven forecasting improves accuracy in wind energy predictions, which is crucial for efficient grid management.

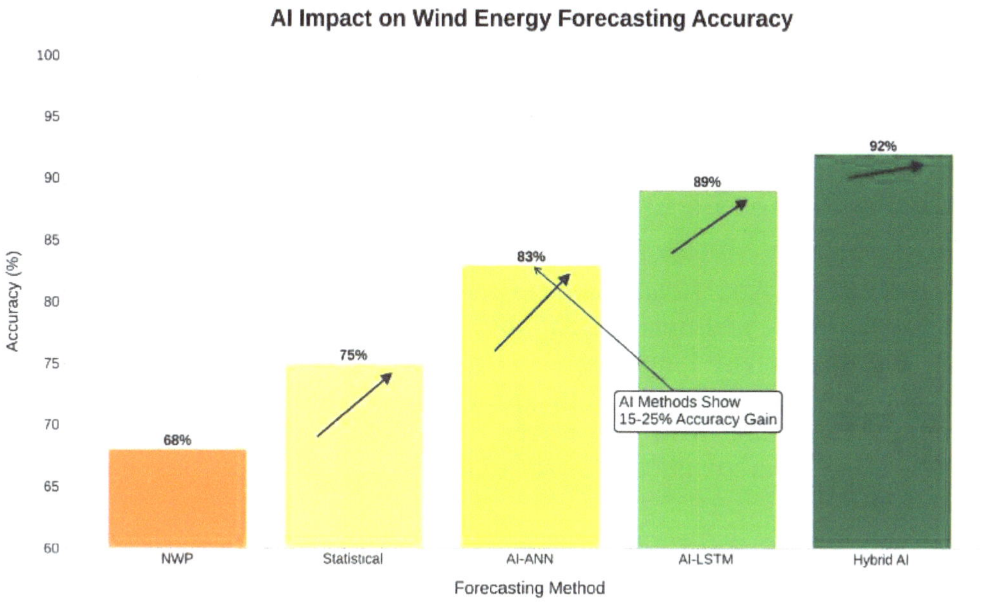

Fig. (6). AI impact on wind energy forecasting accuracy.

- **Machine Learning Models for Wind Energy Forecasting**
    - **Artificial Neural Networks (ANNs):** Identify complex relationships between weather variables and turbine output, enabling more accurate wind power predictions.
    - **Long Short-Term Memory (LSTM):** Ideal for time-series forecasting, capturing dependencies in wind speed fluctuations over different time frames.
    - **Ensemble Learning Models:** Combine multiple algorithms for more robust and adaptive forecasting, reducing errors and improving reliability.
- **Benefits of AI-Powered Forecasting**
    - **Improved Grid Stability:** Better prediction of power fluctuations ensures smooth electricity distribution and reduces blackout risks.
    - **Optimal Energy Storage Utilization:** AI helps balance energy supply by predicting high and low wind periods, enabling better energy storage management.
    - **Reduced Energy Curtailment:** Forecasting ensures that generated energy is efficiently used or stored, minimizing wastage and financial losses.
- **Real-World Example: Google DeepMind's AI Forecasting for Wind Farms**

By optimizing when and how power was distributed to the grid, Google's DeepMind used AI-based forecasting for its wind farms, increasing energy value by 20%. By correctly forecasting wind energy availability 36 hours ahead of time, an AI system increased productivity and income [17].

- **Digital Twin Technology in Wind Energy**

Digital twins are a breakthrough in real-time wind turbine monitoring and optimization. AI-driven digital twins work by:

- **Cloud-Based AI Analytics:** Wind turbine sensor data is sent to cloud-based AI systems that create real-time simulations.
- **Deep Reinforcement Learning (DRL):** AI dynamically adjusts digital twin simulations to optimize power output based on varying wind speeds.
- **Anomaly Detection with Edge AI:** AI detects structural stress, abnormal vibrations, and gearbox issues before failures occur.

A digital twin is a computerized model of a real wind turbine that uses data from the real world to replicate performance in real time. Without requiring physical inspections, operators can detect faults, increase efficiency, and monitor turbine health with AI-powered digital twins [18]. As illustrated in Fig. (7), digital twin technology combined with AI facilitates advanced monitoring and predictive maintenance of wind turbines.

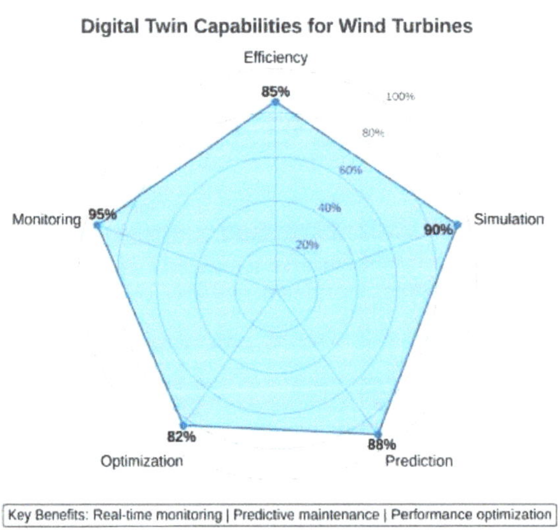

Fig. (7). Shows the role of digital twin technology in wind turbine optimization. AI-powered simulations enhance real-time monitoring, performance prediction, and failure detection.

- **How Digital Twins Work**

    - **Data Collection:** IoT sensors collect real-time data from wind turbines, including wind speed, temperature, vibrations, and power output.
    - **AI-Driven Simulations:** AI models process this data to predict performance trends and potential failures.
    - **Real-Time Optimization:** Operators adjust turbine settings based on digital twin simulations to maximize efficiency and reliability.

- **Key Benefits of Digital Twins**

AI plays a crucial role in enhancing wind turbine operations through several key functionalities. **Failure prediction** enables early detection of mechanical issues before they become serious, thereby minimizing downtime and reducing maintenance costs. **Optimized performance** is achieved through AI-driven simulations that fine-tune blade positioning, yaw control, and load distribution, resulting in improved energy efficiency. Additionally, **remote monitoring and control** capabilities allow operators to oversee turbine performance without frequent on-site inspections, cutting operational expenses and increasing overall reliability. These advancements collectively contribute to more efficient, cost-effective, and resilient wind energy systems. **Case Study: Siemens Gamesa's Digital Twin Technology.** Siemens Gamesa increased turbine lifespan and decreased failure rates by 30% by using digital twins for structural integrity monitoring in offshore wind farms."

- **AI in Wind Farm Asset Management**

AI enhances wind farm performance by implementing advanced asset management strategies that boost efficiency and reduce operational risks. One such strategy is **AI-driven predictive failure analysis**, where real-time turbine data is analyzed against historical failure patterns to anticipate breakdowns before they occur. This proactive approach minimizes unplanned outages and maintenance costs. Additionally, **automated resource allocation** enables maintenance teams to prioritize high-risk turbines, eliminating unnecessary routine inspections and optimizing workforce deployment. Moreover, **multi-agent AI systems** facilitate communication between multiple wind farms, allowing for dynamic energy distribution. This coordination not only minimizes energy waste but also ensures a more balanced and efficient power supply across the grid. By offering real-time analytics, condition monitoring, and predictive maintenance scheduling, AI-driven asset management maximizes wind farm operations. AI provides real-time operational information by enabling centralized monitoring of several wind turbines in sizable wind farms. In order to optimize energy output, it identifies performance irregularities among several turbines and suggests remedial measures [19].

The wind energy industry can reach previously unheard-of levels of efficiency, dependability, and sustainability by incorporating AI into wind turbine optimization, guaranteeing a more stable and affordable renewable energy source in the future [20]. Fig. (**8**) demonstrates the application of AI in optimizing turbine performance, resulting in higher efficiency and reliability.

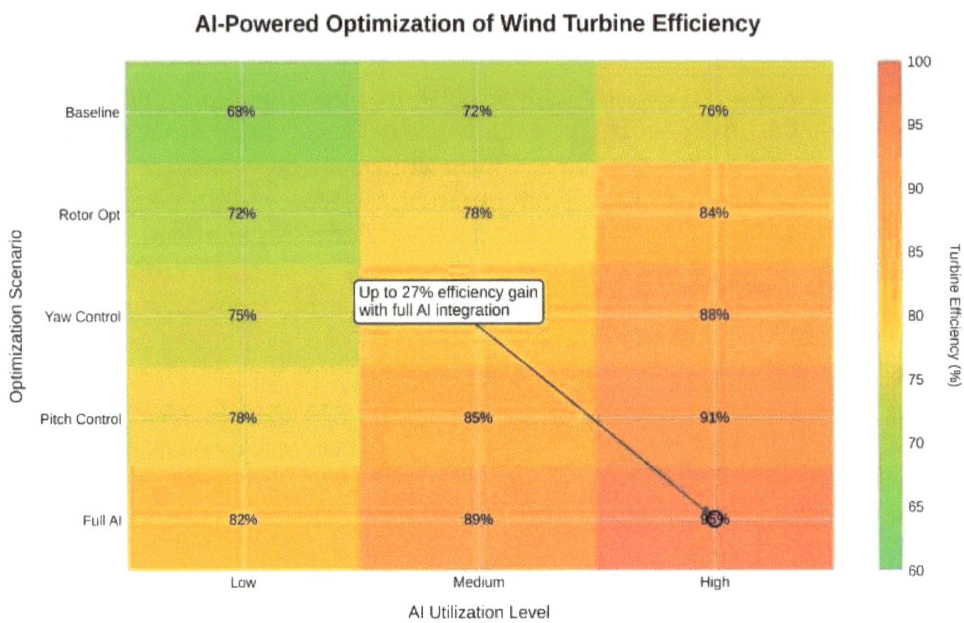

Fig. (8). AI-powered optimization of wind turbine efficiency.

• Benefits of AI-Based Asset Management

AI significantly contributes to the optimization of wind energy systems by addressing key operational areas. It enables **maximized energy production** by detecting inefficiencies in turbine performance and recommending real-time adjustments to enhance power output. In terms of **cost optimization**, AI reduces unnecessary maintenance efforts by prioritizing high-risk turbines, which helps lower overall operational expenses. Furthermore, enhanced **grid integration** is achieved through AI's ability to synchronize wind energy output with power grid demands, minimizing fluctuations and ensuring a more stable and reliable energy supply. These capabilities collectively improve the efficiency, reliability, and cost-effectiveness of wind energy operations. **Case Study: Vestas' AI-Driven**

Wind Farm Optimization. Vestas utilized AI-based asset management software to optimize turbine performance, leading to a 7% increase in annual energy output.

AI-OPTIMIZED WIND TURBINE BLADE DESIGN

Wind turbine blade design is crucial for maximizing energy capture [21]. AI has introduced revolutionary improvements in aerodynamics, material selection, and structural optimization.

- **Traditional *vs.* AI-Based Blade Design**
  a. Traditional Design:
    - Relies on Computational Fluid Dynamics (CFD) and wind tunnel testing.
    - Time-consuming and computationally expensive, requiring multiple iterations.
    - Limited to manual tuning of blade parameters based on expert knowledge.

  b. AI-Driven Design:
    - Uses Generative Adversarial Networks (GANs) to explore millions of blade designs and optimize aerodynamic efficiency.
    - Machine learning models refine blade shape, twist angle, and material composition based on real-world performance data.
    - Deep learning algorithms enhance structural optimization, reducing drag and increasing energy capture.

- **AI for Smart, Self-Adapting Blades**

Future advancements in AI-powered wind turbine technology are set to revolutionize energy efficiency through the development of intelligent, adaptive blades. These smart blades will be capable of dynamically adjusting their shape in real time, responding to changing wind conditions such as speed and direction. By continuously adapting to the environment, they will maintain optimal aerodynamic performance, ensuring maximum energy output and operational efficiency regardless of external fluctuations. This innovation represents a significant leap forward in the evolution of sustainable and intelligent wind energy systems.

Real-World Example: General Electric (GE) AI-Enhanced Blade Design

GE developed AI-optimized blades that improved wind energy conversion efficiency by 15% through smart aerodynamic adjustments and real-time optimizations [22].

- **Future Trends in AI-Driven Wind Turbine Optimization**

  - **AI-Powered Autonomous Wind Turbines**

Future turbines will use self-learning AI models to automatically adjust operations without human intervention.

- **Quantum AI for Ultra-Precise Forecasting**

Quantum computing will allow AI models to simulate millions of weather scenarios instantly, providing near-perfect wind energy predictions.

- **Blockchain-Based AI for Secure Energy Trading**

AI-driven blockchain systems will enhance transparency and efficiency in wind energy transactions, facilitating decentralized energy trading.

- **AI-Integrated 5G and Edge Computing**

5G-enabled AI will process turbine data at the edge, reducing response time and improving efficiency through real-time analytics.

By improving energy forecasting, digital twin simulations, asset management, and blade design, Artificial Intelligence is transforming wind turbine optimization. AI guarantees increased energy efficiency, lower maintenance costs, and increased wind farm reliability by utilizing machine learning, deep learning, and real-time analytics [23].

AI-enhanced blockchain energy trading, autonomous self-learning turbines and quantum AI for extremely precise forecasting are all part of the future of AI-driven wind energy optimization. These developments will strengthen wind energy's position as an affordable and environmentally friendly way to meet the world's electricity needs [24].

CHALLENGES AND FUTURE TRENDS IN AI FOR WIND ENERGY

- **Challenges in AI Adoption**

Even though AI has many benefits for wind energy, a number of obstacles prevent its broad use. These obstacles include everything from technical problems to financial and legal constraints. For AI-driven wind energy systems to reach their full potential, several problems must be resolved [25].

- **High Implementation Costs**

Implementing AI-driven technologies in the wind energy sector, such as predictive maintenance, digital twin systems, and real-time analytics, demands a substantial investment in IoT sensors, cloud infrastructure, and the development of sophisticated AI training models. This financial barrier can be particularly challenging for smaller wind farms, which may face budget limitations that hinder the adoption of such advanced solutions. While the long-term benefits of AI, such as reduced maintenance costs, improved efficiency, and extended equipment lifespan, often outweigh the initial expenses, many operators remain cautious due to the high upfront investment. To accelerate AI integration in wind energy, governments and industry organizations should offer subsidies, incentives, and supportive policies that ease financial burdens and encourage broader adoption of these transformative technologies.

- **Complexity of Data Management**

Wind turbines produce vast volumes of data through IoT sensors that monitor variables such as wind speed, temperature, vibration, and power output. Effectively managing, storing, and analyzing this data demands robust cloud infrastructure, which may not be accessible to all wind farm operators, particularly smaller ones. For AI models to deliver accurate predictions, they require high-quality, well-labeled datasets; however, obtaining consistent and comprehensive historical failure data remains a challenge. The lack of standardized data collection and limited data-sharing practices further hinder model development and scalability. To address these issues, the implementation of **standardized frameworks and collaborative data-sharing mechanisms** is crucial, as they can significantly enhance AI training, improve predictive accuracy, and support broader adoption across the wind energy sector.

- **Cybersecurity Risks**

As wind farms increasingly depend on cloud-based AI analytics and remote monitoring systems, they become more exposed to cybersecurity threats such as hacking, data breaches, and malicious attacks. These vulnerabilities pose serious risks, including the potential for unauthorized turbine shutdowns or disruptions in operations due to compromised AI-driven control systems. To address these concerns, **implementing robust cybersecurity measures**, such as blockchain-based data protection and advanced encryption protocols, is essential. Additionally, **AI-driven anomaly detection systems** can play a critical role in identifying unusual network activity, allowing operators to detect and respond to threats proactively. Strengthening cybersecurity is vital to ensure the safe and uninterrupted operation of AI-integrated wind energy infrastructure.

- **Lack of Standardized AI Frameworks**

The adoption of AI in the wind energy industry is currently hindered by the absence of standardized implementation guidelines, resulting in inconsistent model performance and efficiency across different projects. Many wind farms rely on proprietary AI solutions, which limits the ability to share insights, data, and best practices industry-wide. This fragmentation poses a challenge to achieving large-scale improvements in wind energy operations. To overcome these issues, **regulatory bodies must develop and enforce unified AI adoption frameworks** that promote interoperability among various technologies and systems. Furthermore, **collaboration among governments, research institutions, and the private sector** is essential to establish comprehensive standards that guide responsible and effective AI integration, ensuring scalable, secure, and consistent performance across the wind energy sector.

- **Resistance to AI Adoption**

The integration of AI in wind energy operations can face resistance from human operators and maintenance teams, often due to concerns about job displacement and loss of control over decision-making processes. To address this challenge, it is essential to educate personnel on the advantages of AI-assisted decision-making, emphasizing that AI serves as a tool to enhance, not replace, human expertise. Upskilling technicians to work alongside AI systems through targeted training programs and hands-on experience can foster confidence and competence. By positioning AI as a decision-support system that complements human judgment, organizations can ease the transition, promote collaboration between humans and machines, and ensure a smoother and more widely accepted implementation [26]. Table **3** outlines the primary challenges hindering AI adoption in wind energy and suggests possible strategies to overcome these obstacles.

Table 3. Challenges in AI adoption for wind energy.

Challenge	Description	Potential Solutions
High Implementation Costs	AI-based predictive maintenance, digital twins, and real-time analytics require significant investments in IoT sensors, cloud computing, and AI models.	Government subsidies, long-term cost-benefit analysis, and scalable AI solutions.

(Table 3) cont.....

Challenge	Description	Potential Solutions
Complexity of Data Management	Large amounts of turbine data require advanced storage, processing, and integration capabilities.	Improved data pipelines, AI-driven data structuring, and standardized frameworks.
Cybersecurity Risks	AI-driven wind farms are vulnerable to cyberattacks and hacking attempts.	Implementing blockchain security, AI-based anomaly detection, and robust encryption protocols.
Lack of Standardized AI Frameworks	Different wind farms use proprietary AI solutions, making interoperability difficult.	Regulatory bodies must establish industry-wide AI adoption standards.
Resistance to AI Adoption	Concerns over job displacement and trust in AI-driven decision-making.	Workforce upskilling, AI-assisted (not AI-replaced) decision-making, and training programs.

FUTURE TRENDS IN AI FOR WIND ENERGY

The future of wind energy operations will be shaped by a number of new trends as AI technology develops. Beyond what is now possible, these advances seek to increase automation, boost real-time decision-making, and optimize wind farm performance.

• AI-Powered Autonomous Wind Turbines

The future of wind energy is set to be increasingly autonomous, driven by AI-powered innovations that minimize the need for human intervention. Upcoming wind turbines will be equipped with **self-learning control systems** capable of independently adjusting blade pitch, yaw, and rotation speed in response to real-time wind conditions, ensuring optimal energy output at all times. Additionally, **autonomous drones** will take over inspection and maintenance tasks, significantly reducing the reliance on human field technicians and enhancing safety. Complementing these systems, **AI-powered sensors** will continuously monitor turbine performance, detecting subtle changes and making instant adjustments to maintain peak efficiency. Together, these advancements will lead to smarter, more efficient, and highly autonomous wind energy operations.

• Quantum AI for Ultra-Precise Wind Energy Forecasting

Quantum computing is poised to transform wind energy operations by vastly enhancing the capabilities of AI models. With the ability to simulate millions of weather scenarios almost instantly, quantum computing will significantly improve forecasting accuracy, allowing wind farms to fine-tune power distribution and grid integration with near-perfect precision. Through quantum machine learning algorithms, real-time analysis of complex weather patterns will become possible, leading to smarter decision-making and greater operational efficiency. This advancement not only boosts wind farm profitability but also enables quantum-enhanced predictive maintenance, allowing operators to anticipate turbine failures with exceptional accuracy and take preventive action well in advance. Overall, quantum computing represents the next frontier in making wind energy smarter, more reliable, and highly optimized [27].

- **AI-Driven Blockchain for Secure Energy Transactions**

AI-integrated blockchain platforms are set to revolutionize energy trading by providing a secure and transparent framework for transactions between wind farms and grid operators. These platforms will facilitate decentralized peer-to-peer (P2P) energy markets, enabling individuals and businesses to buy or sell excess wind energy directly without intermediaries. Through AI-powered smart contracts, energy trading processes can be automated, significantly reducing transaction costs and enhancing grid stability by ensuring timely and accurate exchanges. Moreover, blockchain's inherent security features will bolster cybersecurity by preventing unauthorized access and detecting fraudulent activities, thereby fostering trust and integrity within the wind energy market.

- **Edge AI and 5G Integration for Real-Time Optimization**

The integration of Edge AI with 5G networks promises to significantly enhance wind turbine operations by enabling data processing directly at the turbine site, thereby reducing latency and enabling faster, real-time decision-making. Unlike traditional reliance on centralized cloud computing, Edge AI models empower turbines to autonomously adjust their functions based on immediate wind conditions. The high-speed connectivity provided by 5G technology improves communication between wind farms and control centers, facilitating seamless data exchange. Additionally, edge computing supports rapid fault detection and response, which minimizes downtime and boosts overall turbine efficiency. Together, these technologies pave the way for smarter, more responsive, and highly efficient wind energy systems.

- **AI-Optimized Floating Wind Turbines**

Offshore floating wind turbines encounter distinct challenges due to harsh marine environments and highly variable weather conditions. AI technologies are set to enhance these systems by improving stability control mechanisms, which help reduce structural stress and lower maintenance expenses. Advanced Deep

Reinforcement Learning (DRL) algorithms will enable dynamic adjustments of floating platform positions, optimizing wind capture efficiency in real time. Additionally, AI-driven breakthroughs in material science will pave the way for self-repairing turbine blades that are more resistant to corrosion and fatigue, significantly extending their operational lifespan and reliability in demanding offshore conditions [28].

- **Self-Healing AI-Enabled Wind Turbines**

Future wind turbines will incorporate self-repairing AI algorithms capable of identifying and addressing minor issues autonomously, minimizing the need for human intervention. AI-powered nanotechnology-based coatings will protect blades from erosion, significantly extending their operational lifespan. Through predictive analytics, maintenance schedules will be automatically activated as needed, preventing major failures before they occur. Furthermore, advanced AI systems will utilize machine learning to analyze stress patterns in turbine components, driving innovations that enhance overall structural durability and reliability.

- **AI-Integrated Hybrid Renewable Energy Systems**

Future wind farms will function as integral components of AI-optimized hybrid renewable energy systems, seamlessly combining wind, solar, and battery storage technologies. AI models will intelligently and dynamically distribute energy across these diverse sources, continuously adjusting to real-time weather conditions and grid demand. This integrated approach will maximize the overall efficiency, reliability, and sustainability of the renewable energy supply [29].

- **AI for Real-Time Weather Adaptation and Control**

AI will significantly improve wind turbine adaptability by continuously adjusting blade pitch, yaw, and rotational speed in response to real-time weather changes. Through AI-driven wind field estimation, turbines can anticipate gusts before they reach the blades, minimizing mechanical stress and boosting overall efficiency. For example, AI systems can forecast extreme weather events and proactively modify turbine settings to safeguard equipment, preventing damage during storms and ensuring uninterrupted operation.

- **Generative AI for Wind Turbine Design**

Generative AI is set to revolutionize wind turbine development by designing innovative aerodynamic blades that maximize wind capture efficiency. By leveraging AI-driven simulations, engineers can create and test thousands of virtual prototypes, drastically cutting down research and development costs and time before producing physical models. For instance, DeepMind's AI-powered engineering tools have the potential to develop next-generation turbine blades featuring self-adjusting aerodynamic properties, enhancing performance across varying wind conditions.

- **AI in Microgrid and Smart Grid Management**

AI will be essential for real-time grid optimization, facilitating the stable integration of wind power alongside other renewable energy sources. Advanced machine learning algorithms will continuously analyze energy consumption patterns and dynamically adjust distribution, reducing waste and enhancing overall efficiency. For example, AI-driven demand-response management systems can automatically modulate wind power output in response to peak and off-peak grid demands, ensuring a balanced and reliable energy supply.

- **AI-Enhanced Floating Wind Turbines for Offshore Energy**

AI-powered stability control systems are set to enhance the performance of floating wind turbines by enabling them to operate more efficiently in challenging offshore environments. Utilizing Deep Reinforcement Learning (DRL) models, these systems can optimize mooring configurations and dynamically adjust platform positioning to maximize wind energy capture. For instance, AI-based ocean current forecasting can predict changes in water flow and allow turbines to be repositioned in real time, improving both stability and energy output under varying sea conditions.

- **AI-Enabled Predictive Logistics for Maintenance and Supply Chain Optimization**

AI can play a vital role in streamlining the supply chain and maintenance logistics of wind energy operations. By accurately predicting the need for spare parts and repair resources before failures happen, AI ensures timely availability and reduces downtime. AI-powered fleet management systems will monitor turbine health and automatically schedule maintenance activities, improving operational efficiency. For example, wind farm operators can leverage AI-driven logistics planning to strategically deploy technicians where they are most needed, cutting down travel time and significantly lowering operational costs. Table **4** highlights promising future AI trends poised to transform wind energy optimization and management.

Table 4. Future trends in AI for wind energy.

Future Trend	Description	Expected Impact
AI-Powered Autonomous Wind Turbines	AI-driven self-learning control systems dynamically adjust blade pitch, yaw, and speed.	Reduces human intervention, optimizes power generation.
Quantum AI for Energy Forecasting	Quantum computing will allow AI to simulate millions of weather patterns for ultra-precise forecasting.	Improves grid integration and energy reliability.
Blockchain for Energy Trading	AI-powered smart contracts and decentralized platforms enable secure peer-to-peer (P2P) energy transactions.	Enhances transparency, automates renewable energy trading.
Edge AI and 5G Integration	AI models will process wind turbine data locally using Edge AI and 5G networks, reducing reliance on cloud computing.	Enables real-time turbine optimization and faster fault detection.
Self-Healing Wind Turbines	AI-driven control systems will enable turbines to detect and repair minor issues autonomously.	Reduces maintenance costs, extends turbine lifespan.

Fig. (**9**) presents the key future trends in AI adoption that are expected to drive innovation in wind energy technologies.

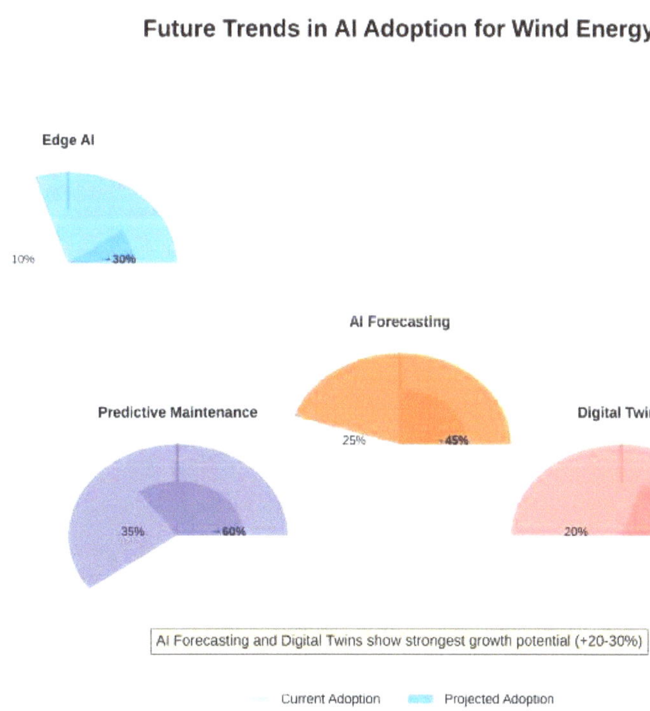

Fig. (9). Future trends in AI adoption for wind energy.

CASE STUDIES ON FUTURE AI APPLICATIONS IN WIND ENERGY

Several leading companies have successfully implemented AI in wind energy operations, showcasing its transformative potential:

Google's DeepMind applied AI-based forecasting to its wind farms, resulting in a 20% increase in energy value. The system provided 36-hour advance predictions, significantly improving power grid integration and enabling smarter energy dispatch planning. Siemens Gamesa utilized AI-driven predictive maintenance, cutting unplanned downtime by 30%. By analyzing vibration and temperature data through machine learning, the company could detect failures early. Additionally, digital twin technology was used to simulate turbine performance and drive operational improvements. Vestas developed AI-powered digital twins to simulate turbine behavior under various conditions. This allowed for pre-deployment testing of performance scenarios, optimization of blade design and yaw control, and continuous learning from operational data to enhance efficiency and maintenance strategies. Ørsted, a leader in offshore wind energy, adopted AI-powered analytics to optimize offshore wind farms. AI improved wave prediction accuracy, enabling better maintenance scheduling and allowing real-time power output adjustments, leading to a 15% boost in energy efficiency. Shell, in partnership with IBM, developed AI tools to forecast failures in offshore wind turbines. Using IBM's Watson AI to process historical and real-time data, the collaboration achieved a 25% reduction in maintenance costs and a marked increase in turbine uptime.

These case studies highlight how AI is revolutionizing wind energy through predictive maintenance, real-time forecasting, and operational optimization.

FUTURE IMPACT OF AI IN WIND ENERGY

By increasing efficiency, cutting prices, and boosting the dependability of renewable energy, AI integration in wind energy is poised to revolutionize the sector. The following are some long-term advantages of using AI [30]:

The future of wind energy is poised for transformation through AI integration across multiple dimensions. AI-driven optimizations are expected to significantly reduce the Levelized Cost of Energy (LCOE), making wind power more economically competitive. Enhanced grid stability and seamless renewable integration will be achieved through advanced forecasting, ensuring reliable energy delivery. The emergence of autonomous wind farms is capable of self-monitoring, operation, and optimization, and will minimize the need for continuous human intervention. Furthermore, AI will drive the development of sustainable hybrid energy systems, integrating wind, solar, and energy storage for greater reliability and efficiency. Importantly, this AI revolution will also reshape the workforce, creating new roles in areas like data science, AI engineering, and predictive analytics, contributing to a more skilled and future-ready renewable energy sector.

With continued advancements in AI, wind energy will become more efficient, resilient, and a cornerstone of the global transition to clean energy.

REGULATORY AND POLICY CONSIDERATIONS FOR AI IN WIND ENERGY

Frameworks for regulations and policies must change when AI is incorporated more deeply into wind energy systems in order to guarantee its safe, open, and effective use. Even though AI has revolutionary potential, issues with ethics, standardization, data protection, and compliance must be resolved.

The integration of AI into wind energy systems brings immense potential but also underscores the pressing need for global governance and ethical oversight. One major challenge is the lack of standardized AI policies, as different countries operate under varying levels of regulation, leading to fragmented adoption and implementation. For instance, while the EU's AI Act seeks to establish comprehensive rules, many regions remain without similar frameworks. Additionally, compliance and data privacy issues arise due to the vast amount of data collected from AI-enabled turbines, necessitating strict regulations like GDPR, though equivalents are missing in several parts of the world. Ethical concerns further complicate matters, especially regarding accountability in AI-driven decisions, raising questions about who is responsible for failures: developers, operators, or manufacturers. To ensure fairness and transparency, AI systems in wind energy must be explainable and auditable, avoiding opaque "black-box" models. Governments must not only introduce clear policies and incentives but also collaborate internationally through bodies like IRENA and the UN to establish unified standards, enabling safe, ethical, and effective AI integration in the global wind energy sector. For example: The European Union's AI [29] Act aims to create AI regulations, but similar frameworks are lacking in many other countries.

CONCLUSION

By greatly increasing predictive maintenance, maximizing turbine performance, and strengthening energy forecasts, Artificial Intelligence (AI) is changing the wind energy environment. Wind farms may increase productivity, lower operating costs, and guarantee a more dependable and sustainable electricity supply by utilizing AI-driven solutions.

AI is still transforming the wind energy industry in spite of significant obstacles like high implementation costs, cybersecurity flaws, and difficult data management. The next generation of wind energy breakthroughs is expected to be driven by developments in edge AI integration, blockchain-enabled energy trading, quantum AI forecasting, and AI-powered autonomous wind turbines. In addition to increasing wind farm efficiency, these advancements will promote increased grid integration and stability, increasing the dependability of renewable energy.

Wind energy will grow in affordability, scalability, and environmental sustainability as AI technology advances. Wind farms will be able to function with greater automation and intelligence through the use of AI-driven technologies, which will decrease downtime and increase turbine lifespans. Furthermore, by enabling intelligent, hybrid renewable energy systems that seamlessly integrate solar, wind, and energy storage technologies, Artificial Intelligence will be a key component of the global shift toward cleaner, greener energy.

The wind energy sector may achieve previously unheard-of levels of creativity, efficiency, and sustainability by fully embracing AI, opening the door to a future driven by intelligent, data-driven renewable energy solutions.

REFERENCES

[1] X. Chen, M.A. Eder, A. Shihavuddin, and D. Zheng, "A Human-Cyber-Physical System toward Intelligent Wind Turbine Operation and Maintenance", *Sustainability (Basel),* vol. 13, no. 2, p. 561, 2021.
[http://dx.doi.org/10.3390/su13020561]

[2] W. Shin, J. Han, and W. Rhee, "AI-assistance for predictive maintenance of renewable energy systems", *Energy,* vol. 221, p. 119775, 2021.
[http://dx.doi.org/10.1016/j.energy.2021.119775]

[3] J. Chatterjee, and N. Dethlefs, "Scientometric review of Artificial Intelligence for operations & maintenance of wind turbines: The past, present and future", *Renew. Sustain. Energy Rev.,* vol. 144, p. 111051, 2021.
[http://dx.doi.org/10.1016/j.rser.2021.111051]

[4] M.S. Mathew, S.T. Kandukuri, and C.W. Omlin, "Estimation of Wind Turbine Performance Degradation with Deep Neural Networks", *PHM Society European Conference,* vol. vol. 7, pp. 351-359, 2022.
[http://dx.doi.org/10.36001/phme.2022.v7i1.3328]

[5] S.Y. Oh, C. Joung, S. Lee, Y-B. Shim, D. Lee, G-E. Cho, J. Jang, I.Y. Lee, and Y-B. Park, "Condition-based maintenance of wind turbine structures: A state-of-the-art review", *Renew. Sustain. Energy Rev.,* vol. 204, p. 114799, 2024.

[http://dx.doi.org/10.1016/j.rser.2024.114799]

[6] D. B. Hulwan, S. Chitra, A. Chokkalingan, K. K. Dixit, S. Subburam, V. Katarwar, and S. Tharmar, "AI-Based Fault Detection and Predictive Maintenance in Wind Power Conversion Systems", In: *Proceedings of the 2024 International Conference on Renewable Energy Resources and Applications (ICRERA-2024), E3S Web Conf.*. vol. 591, Art. no. 02003, Nov. 2024. [http://dx.doi.org/10.1051/e3sconf/202459102003]

[7] A. Kamisetty, "AI-driven robotics in solar and wind energy maintenance: A path toward sustainability", *Asia Pac. J. Energy Environ,* vol. 9, no. 2, pp. 119-128, 2022. [http://dx.doi.org/10.18034/apjee.v9i2.784]

[8] N.O. Farrar, M.H. Ali, and D. Dasgupta, "Artificial intelligence and machine learning in grid connected wind turbine control systems: a comprehensive review", *Energies,* vol. 16, no. 3, p. 1530, 2023. [http://dx.doi.org/10.3390/en16031530]

[9] S. Koukoura, J. Carroll, and A. McDonald, "On the use of AI based vibration condition monitoring of wind turbine gearboxes", *J. Phys. Conf. Ser.,* vol. 1222, no. 1, p. 012045, 2019. [http://dx.doi.org/10.1088/1742-6596/1222/1/012045]

[10] F.P. García Márquez, and A. Peinado Gonzalo, "A comprehensive review of artificial intelligence and wind energy", *Arch. Comput. Methods Eng.,* vol. 29, no. 5, pp. 2935-2958, 2022. [http://dx.doi.org/10.1007/s11831-021-09678-4]

[11] E. Mammadov, M. Farrokhabadi, and C.A. Canizares, *"AI-enabled Predictive Maintenance of Wind Generators," in 2021 IEEE PES Innovative Smart Grid Technologies Europe (ISGT Europe).* IEEE, 2021, pp. 1-5. [http://dx.doi.org/10.1109/ISGTEurope52324.2021.9640162]

[12] D. Astolfi, F. De Caro, and A. Vaccaro, "Condition monitoring of wind turbine systems by explainable artificial intelligence techniques", *Sensors (Basel),* vol. 23, no. 12, p. 5376, 2023. [http://dx.doi.org/10.3390/s23125376] [PMID: 37420542]

[13] Y. Vidal, "Artificial intelligence for wind turbine condition monitoring", *Energies,* vol. 16, no. 4, p. 1632, 2023. [http://dx.doi.org/10.3390/en16041632]

[14] M. Rezamand, M. Kordestani, R. Carriveau, D.S.K. Ting, M.E. Orchard, and M. Saif, "Critical Wind Turbine Components Prognostics: A Comprehensive Review", *IEEE Trans. Instrum. Meas.,* vol. 69, no. 12, pp. 9306-9328, 2020. [http://dx.doi.org/10.1109/TIM.2020.3030165]

[15] A. Keprate, S. Kilskar, and P. Andrews, "Towards Efficient Operation and Maintenance of Wind Farms: Leveraging AI for Minimizing Human Error", In: *Proceedings of the 8th European Conference of the Prognostics and Health Management (PHM) Society,* vol. vol. 8, no. 1, p. 9, 2024. [http://dx.doi.org/10.36001/phme.2024.v8i1.4067]

[16] Y. Merizalde, L. Hernández-Callejo, O. Duque-Perez, and V. Alonso-Gómez, "Maintenance Models Applied to Wind Turbines. A Comprehensive Overview", *Energies,* vol. 12, no. 2, p. 225, 2019. [http://dx.doi.org/10.3390/en12020225]

[17] L. Gong, and Y. Chen, "Machine Learning-enhanced loT and Wireless Sensor Networks for predictive analysis and maintenance in wind turbine systems", *Int. J. Intell. Netw,* vol. 5, pp. 133-144, 2024. [http://dx.doi.org/10.1016/j.ijin.2024.02.002]

[18] X. Jiang, W. Zeng, P. Zhao, L. Cui, and M. Lu, "AI-Based Structural Defects Detection of Offshore Wind Turbines for Automatical Operation and Maintenance," In: *ASME 2023 5th International Offshore Wind Technical Conference, American Society of Mechanical Engineers*, Dec. 2023. [http://dx.doi.org/10.1115/IOWTC2023-119330]

[19] Y.S. Afridi, K. Ahmad, and L. Hassan, "Artificial intelligence based prognostic maintenance of

renewable energy systems: A review of techniques, challenges, and future research directions", *Int. J. Energy Res.*, vol. 46, no. 15, pp. 21619-21642, 2022.
[http://dx.doi.org/10.1002/er.7100]

[20] A. Bielecki, and M. Wójcik, "Hybrid AI system based on ART neural network and Mixture of Gaussians modules with application to intelligent monitoring of the wind turbine", *Appl. Soft Comput.*, vol. 108, p. 107400, 2021.
[http://dx.doi.org/10.1016/j.asoc.2021.107400]

[21] A. Reddy, V. Indragandhi, L. Ravi, and V. Subramaniyaswamy, "Detection of Cracks and damage in wind turbine blades using artificial intelligence-based image analytics", *Measurement,* vol. 147, p. 106823, 2019.
[http://dx.doi.org/10.1016/j.measurement.2019.07.051]

[22] J. Maldonado-Correa, S. Martín-Martínez, E. Artigao, and E. Gómez-Lázaro, "Using SCADA data for wind turbine condition monitoring: a systematic literature review", *Energies,* vol. 13, no. 12, p. 3132, 2020.
[http://dx.doi.org/10.3390/en13123132]

[23] A.A. Noman, Z. Tasneem, M.F. Sahed, S.M. Muyeen, S.K. Das, and F. Alam, "Towards next generation Savonius wind turbine: Artificial intelligence in blade design trends and framework", *Renew. Sustain. Energy Rev.*, vol. 168, p. 112531, 2022.
[http://dx.doi.org/10.1016/j.rser.2022.112531]

[24] A. Peinado Gonzalo, T. Benmessaoud, M. Entezami, and F.P. García Márquez, "Optimal maintenance management of offshore wind turbines by minimizing the costs", *Sustain. Energy Technol. Assess.*, vol. 52, p. 102230, 2022.
[http://dx.doi.org/10.1016/j.seta.2022.102230]

[25] K.S. Wang, V.S. Sharma, and Z.Y. Zhang, "SCADA data based condition monitoring of wind turbines", *Adv. Manuf.*, vol. 2, no. 1, pp. 61-69, 2014.
[http://dx.doi.org/10.1007/s40436-014-0067-0]

[26] A. Aminzadeh, M. Dimitrova, M.S. Meiabadi, S. Sattarpanah Karganroudi, H. Taheri, H. Ibrahim, and Y. Wen, "Non-contact inspection methods for wind turbine blade maintenance: techno–economic review of techniques for integration with industry 4.0", *J. Nondestr. Eval.*, vol. 42, no. 2, p. 54, 2023.
[http://dx.doi.org/10.1007/s10921-023-00967-5]

[27] Y.K. Wu, C.Y. Lee, C.R. Chen, K.W. Hsu, and H.T. Tseng, "Optimization of the Wind Turbine Layout and Transmission System Planning for a Large-Scale Offshore WindFarm by AI Technology", *IEEE Trans. Ind. Appl.*, vol. 50, no. 3, pp. 2071-2080, 2014.
[http://dx.doi.org/10.1109/TIA.2013.2283219]

[28] M.A. Lutz, B. Schäfermeier, R. Sexton, M. Sharp, A. Dima, S. Faulstich, and J.M. Aluri, "KPI extraction from maintenance work orders—a comparison of expert labeling, text classification and AI-assisted tagging for computing failure rates of wind turbines", *Energies,* vol. 16, no. 24, p. 7937, 2023.
[http://dx.doi.org/10.3390/en16247937]

[29] M. A, D. Bisht, S. Yuvaraj, M. Udayapal Reedy, K. Al-Attabi, and A. Dhablia, "AI-Controlled Wind Turbine Systems: Integrating IoT and Machine Learning for Smart Grids," E3S Web of Conferences, vol. 540, p. 03008, Jun. 2024.
[http://dx.doi.org/10.1051/e3sconf/202454003008]

[30] C. Yang, J. Jia, K. He, L. Xue, C. Jiang, S. Liu, B. Zhao, M. Wu, and H. Cui, "Comprehensive analysis and evaluation of the operation and maintenance of offshore wind power systems: a survey", *Energies,* vol. 16, no. 14, p. 5562, 2023.
[http://dx.doi.org/10.3390/en16145562]

CHAPTER 8

Hybrid Wind-Solar Systems Powered by AI

Himani Uppal[1], **Vandita Nandal**[1], **Sheetal**[1,*] and **Inam Ul Haq**[2]

[1] *Department of Computer Science and Engineering (CSE), Shree Guru Gobind Singh Tricentenary University (SGT) University, Gurugram, Haryana, India*

[2] *School of Engineering and Technology (SET), CGC University Mohali, Punjab, India*

Abstract: AI developments promoted the integration of AI into hybrid solar-wind power systems, which has developed to the extent that it introduced a lot of improvement in the field of renewable-prediction accuracy, energy resource allocation, and overall system efficiency. Energy management technologies that constituted the hybrid framework relied on static models and heuristic control methods, which lacked computational efficiency. Conversely, AI-powered approaches leverage Machine Learning (ML) and Deep Learning (DL) algorithms for predictive analytics, smart decision-making, and adaptive energy management. This paper expounds on why AI-driven hybrid energy is superior compared to the previous generation of hybrid systems and highlights some key optimization areas, including energy storage, operational costs, and grid stability, all enhanced by AI. AI-supported sustainable measures will also assist in lowering carbon emissions, safeguarding the environment, and thereby providing greater economic benefits from renewable energy investments. Challenges regarding data availability, computational needs, and cybersecurity continue to affect AI hybrid systems. Such innovations, through improved applications of AI with explainable AI (XAI) and cloud-based energy management, would, among other benefits, ensure the large-scale feasibility of AI hybrid solar-wind systems. The results clearly demonstrate and place AI among the most significant agents in transforming renewable energy management, consequently giving rise to a more transparent and resilient energy future.

Keywords: AI developments, AI-driven decision-making, AI-powered forecasting, Deep learning, Energy resource allocation, Energy storage optimization, Explainable AI (XAI), Grid stability, Hybrid solar-wind systems, Prediction accuracy, Renewable energy optimization, Heuristic control, Machine learning, Sustainability.

* **Corresponding author Sheetal:** Department of Computer Science and Engineering (CSE), Shree Guru Gobind Singh Tricentenary University (SGT) University, Gurugram, Haryana, India; E-mail: sheetal_feat@sgtuniversity.org

Sugandha Singh, Inam Ul Haq, Sachin Ahuja, Akib Mohi Ud Din Khanday & Neha Sharma (Eds.)

INTRODUCTION

The Rising Need for Sustainable Energy

The demand for sustainable energy is increasing globally, and traditional fossil fuels are vanishing, leading to environmental issues. Hybrid wind-solar systems are emerging as a powerful solution to meet this demand. Trends over the past several years show that the prospects of energy resource development have played a very important role in the development of human society. Since the Industrial Revolution, energy has been the wheel of development for modern civilization. The other factors were technological development and energy consumption, whereas the third factor was the increasing population of the world [1]. Ensuring a long-term energy supply as traditional sources are finite means transitioning to renewable energy sources like solar, wind, and hydropower that can be replenished naturally, ultimately leading to an eco-friendly

and a sustainable future. In the 21st century, the world faces an ever-growing energy demand, driven by rapid industrialization, urban expansion, and the increasing electrification of daily life. However, the traditional energy sources that have powered human civilization for centuries, fossil fuels, are finite and pose significant environmental threats. The need for sustainable energy solutions has never been more critical, as the consequences of climate change, air pollution, and resource depletion continue to threaten ecosystems and economies alike [2].

The Environmental and Economic Consequences of Fossil Fuels

The process of burning fossil fuels releases large amounts of greenhouse gases, like carbon dioxide (CO_2). These gases absorb heat and, as a consequence, contribute to global warming. Intensifying temperatures have led to extreme weather patterns, rising sea levels, and disruptions to agriculture and biodiversity. Moreover, reliance on fossil fuels has geopolitical and economic implications, as many countries depend on energy imports, making them vulnerable to price fluctuations and supply chain disruptions [3].

Beyond environmental consequences, reliance on fossil fuels has significant geopolitical and economic implications. Many countries depend on energy imports, making them vulnerable to price fluctuations and supply chain disruptions. Political instability in oil-rich regions, trade conflicts, and global economic crises have historically led to sharp variations in fuel prices, affecting both consumers and industries. Fig. (**1**) outlines the major environmental and economic consequences associated with fossil fuel consumption.

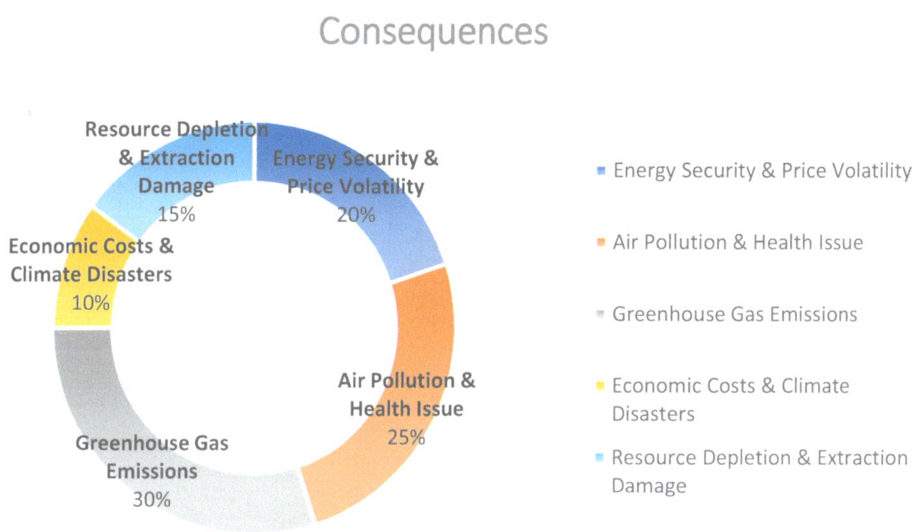

Fig. (1). The environmental and economic consequences of fossil fuels.

The Shift Towards Renewable Energy

A sustainable substitute for fossil fuels is provided by renewable energy sources like geothermal, hydro, solar, and wind. These energy sources may be used endlessly, are plentiful, and emit very little. To move toward a more sustainable and independent energy system, governments and businesses worldwide are making significant investments in renewable energy projects. However, despite their advantages, renewable energy systems face challenges related to intermittency, efficiency, and storage [4].

Addressing Energy Intermittency with Hybrid Solutions

The main barrier in adapting renewable energy at a large scale is intermittency-solar power generation is limited to daylight hours. On the other hand, wind energy is reliant on wind speeds, which change throughout the day and season. These variations can be lessened with the use of a hybrid energy system that integrates wind and solar electricity, guaranteeing a more steady and dependable power source. By integrating these complementary energy sources, hybrid wind-solar systems maximize energy production [5].

The transition to sustainable energy is no longer an option but a necessity. Hybrid wind-solar systems, empowered by AI, represent a significant step forward in achieving energy security, reducing carbon footprints, and ensuring a cleaner and more resilient future. With the increasing focus of AI on optimizing the uses of renewable energy, the clear finding is that most of the impediments to large-scale adoption of renewable energy could be tackled by technological solutions. Fig. (**2**) illustrates the factors driving the growing global need for sustainable energy solutions.

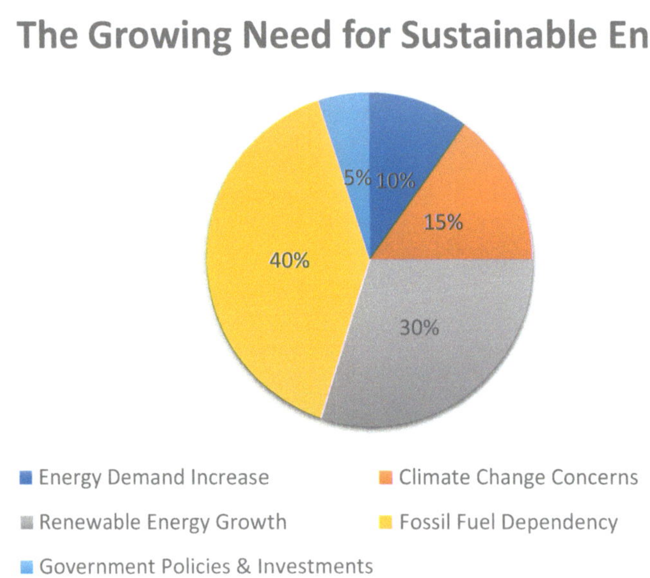

Fig. (2). The growing need for sustainable energy.

WHAT IS A SOLAR ENERGY SYSTEM?

A solar energy system harnesses sunlight to generate electricity through photovoltaic (PV) panels. For usage in homes and businesses, these panels convert solar energy into direct current (DC) electricity, which is subsequently converted into alternating current (AC). However, its efficiency depends on geographical location, weather conditions, and the time of day, posing challenges to consistent energy production [3].

One of the planet's most plentiful and sustainable energy supplies is solar power. It can be transformed into heat or power using various technologies. Solar energy is an essential renewable energy source that helps mitigate environmental problems like climate change and reduces the dependency on fossil fuels.

Harnessing Solar Energy

The energy from the sun can be captured and utilized through different methods: Fig. (**3**) depicts the process of solar energy conversion and distribution.

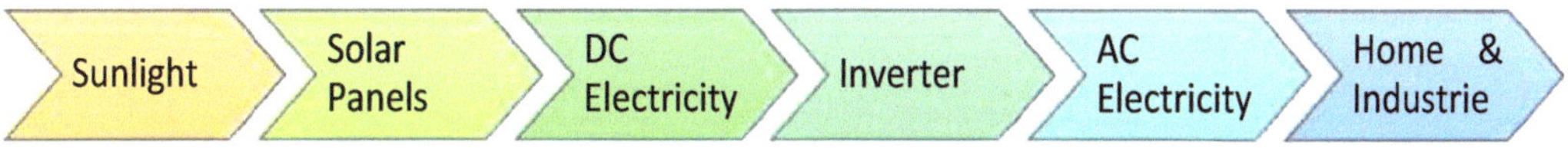

Fig. (3). The flow of solar energy.

- **Solar Photovoltaic (PV) Systems:** These systems use solar panels made of a material named semiconductor that converts sun rays directly into a power supply. PV technology has experienced wide application in residences, commercial buildings, and industrial structures.
- **Solar Thermal Systems:** This method captures solar heat for water heating, space heating, and power generation. To transmit heat to a working fluid for use in a variety of heating applications, solar thermal collectors absorb sunlight.
- **Concentrated Solar Power (CSP):** CSPs use combinations of mirrors or lenses to focus sunlight on a small area. The sunlight focuses on a small area and generates a very high temperature, which in turn produces steam. This steam drives turbines to produce electricity, making CSPs appropriate for developing power plants of large sizes.

Advantages of Solar Energy

- **Renewable and Sustainable:** Solar energy is derived from the sun, which is a non-depletable resource. In contrast to finite fossil fuels, sunlight is available daily and will continue to be a viable source of energy for billions of years. Solar energy is thus a long-term, sustainable, and renewable source of power.
- **Environmentally Friendly:** The largest advantage of solar energy is that it has a light environmental footprint. The production of solar energy is not like the use of fossil fuels, which emit greenhouse gases such as carbon dioxide (CO_2) or methane (CH_4) that cause global warming. Solar energy doesn't cause air or water pollution, because it is a clean source of energy that preserves ecosystems and biodiversity.
- **Low Operating Costs:** Although the expenses of installing solar panels are high initially, the advantages in the long run are several times more than the expenditure. Solar panels are easy to maintain once installed and can function for 25 - 30 years. Solar energy is offered by the sun for free, thus lowering the cost of electricity. Additionally, technological innovation in solar technology and government incentives in the form of credits and subsidies are making solar power affordable and accessible.

- **Energy Independence:** Solar energy minimizes the use of fossil fuels, whose prices are driven by geopolitical tensions, supply disruptions, and price volatility. With solar power, homes, businesses, and even countries can be more energy secure. This independence is especially precious in rural and off-grid locations where access to traditional electricity is too costly or simply not feasible.
- **Scalability and Versatility:** Solar power systems are highly flexible, with projects ranging from residential rooftop systems to gigantic solar farms that can supply energy to entire cities. Solar panels can even be installed in buildings, cars, and portable devices, making solar energy very adaptable for all sorts of applications.
- **Job Creation and Economic Growth:** Rapid growth of the solar industry correlates with increased employment opportunities not only in manufacturing and installation but also in maintenance and research development. Investment in solar energy indeed saves the environment while simultaneously leading to economic growth through job creation and the evolution of new clean energy technologies.

WHAT IS A WIND ENERGY SYSTEM?

The wind creates energy in an environmentally safe and renewable way, and this energy is captured from the current movement of air by the kinetic energy forces generated through natural wind. In other words, wind energy systems convert this natural force to an energy resource- mechanical or electric. Thus, wind energy is one of the most commonly used energy sources among renewable energies in the world today. The increasing popularity of wind energy will also provide solutions to sustainable energy and reduced dependence on fossil fuels.

Working Principle of a Wind Energy System

A wind energy system operates by converting the wind's kinetic energy into usable electricity through the following steps:

- **Wind Turbines Capture Wind Energy:** Large blades mounted on a wind turbine rotate whenever the wind blows. The speed and efficiency of energy capture depend on wind velocity and turbine design.
- **Mechanical Energy Conversion:** The rotational speed is increased by the revolving blades driving a shaft that is attached to a gearbox. In order to generate electricity, this mechanical motion is necessary.
- **Electrical Energy Generation:** The gearbox transfers mechanical energy to a generator, where it is converted into electricity. This electricity is then transmitted through power lines for distribution.

Types of Wind Energy Systems

Wind energy systems are classified into two types depending on their installation and energy generation scale. Fig. (**4**) categorizes the main types of wind energy systems used globally.

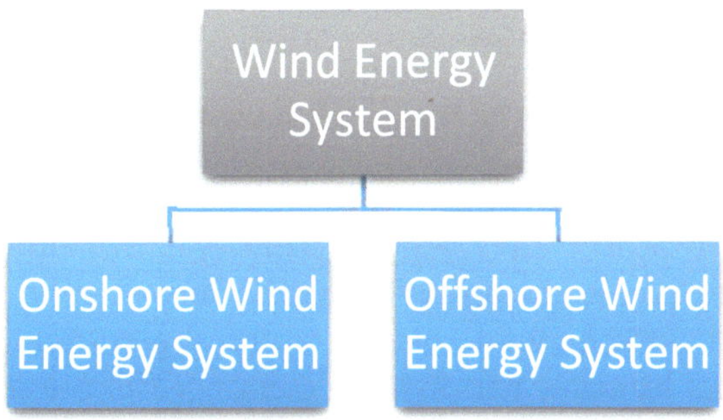

Fig. (4). Types of wind energy systems.

- **Onshore Wind Energy Systems:** Installed on land, these turbines are typically used for large-scale power generation. These turbines are inexpensive but might face issues such as land use and noise associated with their installation.
- **Offshore Wind Energy Systems:** Turbines located in offshore wind farms are characterized by high winds, usually bringing strong, consistent winds. This would result in higher energy output. Their major disadvantage would normally be related to the high installation and maintenance costs.

Advantages of Wind Energy

- **Renewable and Sustainable:** Wind power draws upon the natural movement of air and is hence an infinite, renewable energy. Unlike fossil fuels, which are exhausted over a period of time, wind power is continuously replenished by processes in the atmosphere. This yields a sustainable and long-term method of generating electricity without worrying about depletion.
- **Environmentally Friendly:** The best advantage of wind power is that it does not lead to air pollution or the emission of any greenhouse gas. Wind power generates electricity without burning fossil fuels, therefore reducing the carbon dioxide (CO_2) emissions and other poisonous chemicals that contribute to global warming, smog, and respiratory diseases. Wind power reduces the overall carbon footprint and makes the environment healthier and cleaner.
- **Cost-Effective:** Although the cost of setting up wind farms or home wind turbines is high initially, wind energy is very cheap to run and maintain. Wind turbines require regular maintenance but are capable of running without fuel.

Government subsidies, tax credits, and incentives also lower the cost for mass and individual consumers.

- **Scalable and Versatile:** Wind power systems are highly flexible and can be utilized in various capacities, from small domestic turbines to powerhouses to large wind farms that produce electricity for entire cities. Offshore wind farms experience stronger and more stable winds overseas, resulting in higher efficiency and power generation.

WHAT IS A HYBRID WIND- SOLAR SYSTEM?

A Hybrid Wind-Solar System incorporates wind turbines and solar photovoltaic (PV) panels into a renewable energy generating system that harnesses the power of the sun, giving an ideal and reliable electric power supply. It serves for the combined effect of these renewable energy sources: electrical energy is produced during sunshine hours by solar panels, while energy tilted toward still nights and cloudy days usually flows through wind turbines. By merging both energy sources, hybrid systems can give a more consistent and stable energy output as compared to using either standalone wind or solar systems [6].

How Hybrid Wind-Solar Systems Work

In hybrid systems, the wind turbine uses wind energy to generate electricity along with the solar panel that builds electricity by absorbing sunlight [7]. An inverter then regulates the produced electricity, transforming alternating current (AC) from wind turbines and direct current (DC) from solar panels into a steady power output appropriate for use in residences, workplaces, or industrial facilities [8]. Extra energy is either sent into the grid or saved in battery storage devices for later use.

Advantages of Hybrid Wind-Solar Systems

- **Enhanced Reliability:** Wind and solar sources are thus differentiated, as one will generate power during the day while the other generates at night or in cloudy weather: wind generates at night or in cloudy weather.
- **Reduced Energy Costs:** Hybrid systems contribute very low costs in terms of fossil fuel electricity, and hence, total energy consumption using various renewable sources.
- **Sustainability:** To a great extent, hybrid systems reduce carbon emissions and massively contribute to the efforts of carbon mitigation and climate change efforts all over the globe.
- **Grid Independence:** These systems are ideal for off-grid and remote locations where traditional power infrastructure is unavailable or unreliable.

WHAT IS ARTIFICIAL INTELLIGENCE (AI)?

Artificial intelligence, or AI, is a field of computer science that is associated with machine learning, which carries out tasks that require human intelligence in terms of the operations it performs. Such operations can entail learning, reasoning, problem-solving, perception, language comprehension, as well as decision-making [9]. It is generally divided into general AI, which tries to simulate the human mind in activities varying from one field to another, and narrow AI, which is designated for specialized activities such as voice assistants and recommendation systems.

Key Components of AI

- **Machine Learning (ML):** AI systems that learn through patterns in the data can improve their performance over time without an explicit programming sequence. ML finds applications in the domains of fraud detection and predictive analytics, among many others [10].
- **Deep Learning:** The branch of ML that concerns the neural networks working with massive amounts of data, inferring complex patterns of great importance for such applications as image and speech recognition [11].
- **Natural Language Processing (NLP):** This enables AI to understand and generate human languages, such as chatbots and voice assistants [12].
- **Computer Vision:** AI's ability to transmit and analyze visual data, used in facial recognition, medical imaging, and autonomous vehicles [13].
- **Robotics:** AI-driven robots that perform tasks in industries like manufacturing, healthcare, and space exploration [14].

Artificial intelligence has performed its magic in most sectors, from healthcare to finance, transport, and load dispatching. Its integration into renewable energy systems, like wind-solar hybrid setups, helps optimize energy production, improve efficiency, and reduce costs.

Role of AI in Optimizing Renewable Energy Systems

Increasing efficiency, reliability, and economy was a paramount aim of AI in the backdrop of renewable energy, whereby it has transformed the domain of renewable energy. Key applications of AI in hybrid wind-solar systems include:

- **Smart Energy Forecasting**

AI-powered models analyze historical weather data, real-time meteorological information, and satellite imaging to predict energy production accurately. This allows for better planning and management of energy distribution, minimizing waste and ensuring a consistent power supply [7].

- **Intelligent Energy Management**

AI-driven energy management systems monitor energy consumption and optimize the balance between solar and wind energy generation. These systems dynamically adjust power flow to meet demand while minimizing reliance on non-renewable backup sources [8].

- **Grid Stability and Load Balancing**

AI enables smart grid technologies that can detect and respond to fluctuations in energy supply and demand. By predicting potential shortages or surpluses, AI ensures that electricity distribution remains stable, reducing power outages and grid failures [9].

- **Predictive Maintenance and Fault Detection**

AI enhances the longevity of wind and solar equipment by identifying operational inefficiencies and potential failures before they escalate. AI-driven diagnostics reduce downtime and maintenance costs, ensuring that renewable energy systems operate at peak performance [15].

- **Energy Storage Optimization**

AI algorithms manage battery storage more efficiently by optimizing charging and discharging cycles based on predicted energy supply and demand. This maximizes energy storage capacity, extends battery lifespan, and enhances system reliability [16].

The integration of AI in hybrid wind-solar systems is paving the way for a more efficient and cost-effective energy future. With the advancement of AI technology, its role in renewable energy management will become even more vital, enabling a cleaner and more sustainable world.

TRADITIONAL METHODS USED FOR HYBRID WIND-SOLAR SYSTEMS

Traditional hybrid solar-wind systems methods are basically heuristic in nature, and the common optimization techniques are used for determining configuration, dispatch of energy, and efficiency related to the operation of the systems. All these methods were traditionally developed before Artificial Intelligence (AI) gained popularity in the energy sector and before the advent of machine learning [17].

Introduction to Traditional Hybrid System Optimization

Energy availability and dependability have long been increased through the usage of hybrid energy systems. Deterministic models, rule-based controls, mathematical programming, and heuristic algorithms are examples of conventional hybrid energy optimization techniques [18]. These methods, which did not include adaptive mechanisms or real-time learning, sought to maximize cost, energy efficiency, and resource use [19]. Early hybrid systems were less effective in dynamically changing environmental conditions since they mostly depended on set operation modes [20]. Traditional hybrid system control strategies were primarily based on fixed threshold values and did not incorporate predictive analytics for decision-making [21].

Rule-Based Methods

One of the most widely used traditional approaches for managing hybrid energy systems is rule-based control. These methods depend on predefined logical conditions to switch between energy sources and storage units based on real-time data.

Other than excluding a few exceptions, rule-based strategies are the most popular and simplest methods for managing the hybrid solar-wind energy systems when considering the entire literature [20].

Energy dispatching decisions in rule-based systems function in the following areas:

- Battery charge levels
- Solar radiation intensity
- Wind speed
- Fluctuations in load demand

However, in the area of adaptation and optimization of energy use during unpredictable weather conditions, they simply do not suffice [22]. One of the major drawbacks of rule-based hybrid system control is its inability to optimize energy flows dynamically, leading to frequent energy curtailment and excessive battery cycling [23].

Load Following Strategies (LFS)

LFS entails tweaking the power generation for real-time demand. These techniques use predetermined generation thresholds to determine whether solar/wind or battery storage is employed. Load following control strategies are applied for balancing the supply with the demand [24].

Conditions favourable for LFS methods in order to enhance reliability also suffer from drawbacks:

- Limited storage optimization: Excess energy is wasted beyond storage capacity.
- Slow response to sudden load change: Manual operations are sometimes required.
- Controlled logic is fixed: Not ideal for managing varying wind and solar inputs.

The following methods are effective for standalone hybrid microgrids but fail to integrate with the grid-connected hybrid systems optimally due to their reliance on static thresholds rather than real-time forecasts [25].

Mathematical Optimization Techniques

A few traditional mathematical methods are used to optimize the hybrid system sizing and operations, which are as follows:

- **Linear Programming (LP)**

The LP is mainly the most common method for minimizing costs and dispatching energy. It builds energy generation as an objective function under constraints of demand and resources. Linear programming methods have been widely applied to optimize hybrid system costs, though they struggle with handling nonlinear energy variations [25].

- **Dynamic Programming (DP)**

It is a technique that employs recursive optimization to decide the deterministic storage operations as well as the present generational scheduling.

DP-based hybrid system models were computationally expensive and required significant storage space for multi-stage decision making [26].

- **Mixed-Integer Programming (MIP)**

Mixed Integer Programming (MIP) is a more general case of LP, which has both integer and continuous variables in it. MIP finds its common applications in hybrid system components selection and cost optimization.

MIP approaches were useful in sizing renewable energy components but suffered from extensive computational overhead [27].

Despite the accuracy of mathematical optimization techniques, their application in real-time hybrid system management is constrained by computational complexity and the need for precise input parameters [28].

Probabilistic Methods

Because of its intermittent nature, renewable energy led ultimately to the development of approaches dealing with probability. Among them are:

- Monte Carlo simulation: To simulate solar and wind generation uncertainty.
- Markov Chains: For power fluctuation analysis.
- Statistical Regression Models: Mainly used for forecasting the energy demand.

Monte Carlo simulations offered insights into system reliability but required large datasets for accurate predictions [29]. These models provided valuable insights into energy system reliability but lacked real-time adaptability and often required extensive computation time [30].

Empirical and Heuristic-Based Approaches

Heuristic techniques were traditionally introduced in order to enhance optimization efficiency. They include:

- Genetic Algorithms (GA): Used for sizing the optimal system.
- Particle Swarm Optimization (PSO): Energy dispatch.
- Simulated Annealing (SA): Used to reduce expenses.

Heuristic-based methods such as GA and PSO have been successfully used in hybrid energy optimization but lacked real-time adaptability [27]. However, heuristics required manual fine-tuning and lacked adaptability to sudden energy demand shifts [29]. Genetic algorithms and swarm intelligence approaches have demonstrated promising results in hybrid system sizing, yet they remain computationally intensive for large-scale implementations [30].

CHALLENGES IN TRADITIONAL METHODS OF HYBRID WIND-SOLAR SYSTEMS

- Adaptability to Environmental Change Is Limited: One approach that relies on static thresholds and set subjective parameters, *e.g.*, rule-based control and load-following strategies, is not able to dynamically adjust to the variation in solar irradiation and fluctuation in wind speed or sudden demand variations, thereby severely affecting the overall efficiency of energy management schemes.
- Mathematical Model Computational Complexity: LP and MIP are the resource-hungry tools, and their complexities rise exponentially with the system size, thus shelving any real-time optimization option for large-scale hybrid systems.
- Highly Sensitive to Input Data Accuracy: Probabilistic analysis, like Markov chains and Monte Carlo simulation, heavily depends on historical data. However, a small error in the estimation of the weather forecast value or energy demand might easily throw off the compromise optimization and reliability of the system.
- Inefficient Energy Storage Operations: Admittedly, old methods fare badly in many situations for the optimization of energy storage. In this situation, the load-following strategy may eat up energy, just pushing it over the storage limit; conversely, the alternative rule-based control could be running the battery through too many cycles, which runs down battery capacity.
- Slow Convergence of the Heuristic: Convergence to the optimum takes time with optimization methods such as Genetic Algorithms, Particle Swarm Optimization, and Simulated Annealing. Their slow convergence renders such methods unfit for real-time energy management of dynamic hybrid systems.
- Little Scalability Evidence in Massive Hybrid Systems: Most accepted methodologies have come from small-scale applications. The increasing size and complexity of hybrid energy systems thus appear to be an unveiled challenge against the scalability of the traditional rule-based models and deterministic methods.
- Lack of Predictive Ability: All different forms of traditional models for energy dispatch are reactive systems, seldom applying any predictive modelling. Its absence leads them to be unable to foresee the fluctuations in solar and wind generation. Therefore, at best, their dispatch decisions are sub-optimal.
- Regulatory and Integration Barriers with the Power Grid: The hybrid systems require integration into the prevailing grid, whereby the updating regulatory stipulations, grid-stability requirements, and demand-response mechanisms along the path of system deployment still pose great impediments to the customarily used optimization models.
- Limited Fault Detection of System; Unreliable: Classical hybrid energy management hardly ever uses fault diagnosis methods for real-time fault detection. Unless such a diagnosis is well defined, AI-supported anomaly

detection may fail to detect component failure, voltage fluctuations, and grid instability until close to an impending major crisis.

- Economic Options' Scarcity and Feasibility Problems: Most classical optimization methods aggravate economic uncertainties attached to changes in electricity tariffs, policy changes, and inflation. Hence, this constitutes a serious impediment to long-term financing of hybrid energy systems.

Table **1** compares traditional wind forecasting methods, outlining their key characteristics and limitations.

Table 1. Comparison of different traditional methods.

Methods	Strengths	Weaknesses	Efficiency	Adaptability	Computational Complexity
Rule-Based Control	Simple, easy to implement.	Not adaptive to real-time variations.	Moderate	Low	Low
Load Following Strategies (LFS)	Ensures supply-demand balance	Inefficient power utilization	Moderate	Low	Low
Linear Programming (LP)	Cost minimization, deterministic	Computationally expensive	High	Low	High
Dynamic Programming (DP)	Optimizes energy storage	Complex and time-consuming	High	Moderate	High
Mixed-Integer Programming (MIP)	Component selection optimization	High computational demand	High	Moderate	Very High
Monte Carlo Simulations	Provides reliability assessment	Requires extensive data	Moderate	Low	High
Markov Models	Good for power fluctuation analysis	Slow in real-time applications.	Moderate	Low	High
Genetic Algorithms (GA)	Effective for system sizing	Requires parameter tuning	High	Moderate	High
Particle Swarm Optimization (PSO)	Efficient for energy dispatch	Lacks adaptability to dynamic conditions	High	Moderate	High
Simulated Annealing (SA)	Good for cost minimization	Time-consuming convergence	High	Low	High

INTRODUCTION TO AI IN HYBRID SOLAR-WIND SYSTEMS

The use of Artificial Intelligence (AI) for hybrid solar-wind power has been a huge leap towards efficiency, reliability, and flexibility of the systems.

Optimization processes were previously conducted based on static models and manual controls instead of adopting data-driven decisions in real-time for maximizing production, storage, and delivery of energy using AI [31]. Indeed, the so-called hybrids, machine learning, deep learning, neural networks, fuzzy logic, and evolutionary algorithms, are among the approaches supporting a forward-looking trend in research on hybrid energy systems [32]. Artificial Intelligence has transformed hybrid renewable energy systems by enabling predictive analytics, self-learning algorithms, and adaptive control mechanisms to optimize performance [33]. AI enables automated control methods, dynamic load balancing, and smart energy forecasting, which all improve the efficiency of the system. Different studies have proven that AI models have effectively proven capable of reducing not only energy wastage but also operational costs while increasing grid stability. With the use of AI, hybrid energy systems can now analyse large amounts of meteorological data to enhance forecasting accuracy, thereby ensuring optimal energy generation and utilization [34].

AI-DRIVEN OPTIMIZATION TECHNIQUES

Hybrid energy systems optimization is essential to select the best combination of sources, storage systems, and energy distribution to loads. AI-empowered methods, such as genetic algorithms, particle swarm optimization (PSO), and deep reinforcement learning (DRL), have been effective in optimizing these hybrid systems [35]. Genetic algorithms and swarm intelligence techniques have outperformed traditional rule-based control systems in managing energy distribution in hybrid solar-wind systems [36]. Mainly, these techniques have one purpose: dynamic adjustment of energy generation to the demands of real-time usage and variable weather conditions [37]. This is one of the factors that can contribute to improvement in efficiency [38]. Moreover, hybrid AI strategies have been put out to improve the flexibility and resilience of hybrid solar-wind energy management by combining several optimization methodologies [39]. A hybrid AI framework integrating fuzzy logic, machine learning, and evolutionary algorithms has demonstrated superior performance in optimizing energy distribution and reducing operational costs [40].

AI-Powered Control Systems

Controlling the decision-making in hybrid solar-wind systems through AI-based mechanisms reduces human intervention and helps make the system even more stable. Fuzzy logic controllers and neural networks are largely applied to carry out real-time energy management. Fuzzy logic-based controllers have demonstrated superior adaptability in hybrid energy systems by handling uncertainties in solar and wind energy generation [41].

Neural network-based control models enhance grid stability by predicting power fluctuations and adjusting energy flow accordingly [42]. AI-powered grid management systems are also being developed to enable self-healing smart grids, which can autonomously detect and react to issues. This will ensure a continuous flow of energy. The integration of AI-driven self-healing grid management in hybrid energy systems has resulted in the enhancement of fault detection, faster response times, and enhanced overall system resilience [43].

Case Studies and Real-World Applications

A few case studies highlight the successful implementation of AI in hybrid solar-wind systems across various regions. A hybrid power plant integrating AI for energy management attained an increase in efficiency by a 25% or a 30% reduction in operational costs [36]. AI-driven forecasting models deployed in urban hybrid microgrids have significantly reduced energy wastage and improved grid reliability [38].

Moreover, battery energy storage systems have also been optimized by AI usage to enable hybrid solar-wind plants to optimally store excess amounts of energy for discharge on demand. Machine learning-based battery management systems have enabled predictive maintenance and dynamic charge-discharge scheduling, extending battery lifespan and improving energy availability [44].

MACHINE LEARNING AND DEEP LEARNING APPROACHES IN HYBRID SOLAR-WIND SYSTEMS

With the advent of hybrid solar-wind energy systems, such extreme forecasting, optimization, and decision-making found a relevant space, whereupon machine learning and deep learning algorithms come into play [45]. Traditional or classical methods operated on non-adaptive or fixed models, whereas ML and DL methods have relied on data-driven models to predict energy generation to have a better way of distributing energies and to enhance grid stability. These have been researched and intensively applied on many occasions in hybrid renewable energy systems to enhance performance. Machine learning and deep learning have significantly improved energy management in hybrid solar-wind systems by providing real-time predictive analytics and adaptive optimization techniques [46].

Process involved in ML and DL in Hybrid Systems

The successful hybrid solar-wind system's flawless working involves machine learning and deep learning arranged in multiple phases. The procedures from data collection to model deployment explaining how the most efficient, and accurate energy management system is to be achieved are shown herein.

- Data Collection and Sources

The foundation of any ML or DL-based system is high-quality data. Data collection for hybrid solar-wind systems includes obtaining pertinent parameters like:

- **Meteorological data:** Solar irradiance, wind speed, temperature, humidity, and atmospheric pressure (obtained from satellite imagery, weather stations, IoT sensors, and climate databases).
- **Energy system performance data:** Historical energy generation, battery charge-discharge cycles, and grid demand fluctuations.
- **Geospatial data:** Terrain elevation, land use, proximity to power infrastructure for site selection.

Accurate data collection plays a crucial role in AI-driven renewable energy forecasting, as inconsistencies can lead to poor system performance [47 - 51].

- Data Preprocessing and Feature Engineering

Once data is collected, it must be cleaned and transformed to ensure its quality for further utility. Some important preprocessing includes:

- **Handling missing data:** Data filling methods like mean/mode substitution or deep learning-based interpolation.
- **Noise reduction:** Filtering techniques (*e.g.*, moving average, wavelet transform) to smooth sensor readings.
- **Feature Selection and Extraction:** Identifying the primary variables affecting energy generation while throwing out irrelevant attributes.
- **Normalization and Scaling:** All the parameters are brought into a uniform range to increase stability of the model.

Applying proper preprocessing techniques significantly improves ML model accuracy by reducing redundancy and enhancing critical features [52 - 57].

- Machine Learning and Deep Learning Model Selection

Depending on whether the goal is forecasting, optimization, or decision making, an appropriate selection of models is determined. The selection process can be subdivided into:

- **Supervised Learning Models**
- Regression Models (Linear Regression, Random Forest, XGBoost) for the prediction of energy output.
- Fault detection in renewable grids by Classification Models (SVM, Decision Trees, Naïve Bayes).
- **Deep Learning Models**
- LSTM for Marine forecasting by time series of Wind and Solar power generation.
- CNN to Weather pattern k-analyse through satellite images.
- Transformer Networks with more versatility in time-series forward-looking activities.
- Model Training and Hyperparameter Tuning

Training of the model involves feeding in historical data for it to recognise patterns and eventually make predictions. Some of the important steps are as follows:

- **Splitting dataset:** Splitting data into training (80%) and testing (20%) sets.
- **Optimizing Loss function:** As for predicting highly reliable outcomes, there are afew techniques that are often used: for predicting error, we have the root mean square error, as well as the mean absolute error.
- **Hyperparameter tuning:** Optimization approaches may include using grid search or Bayesian optimization to tune model parameters like learning rate, number of layers, drop-out rate, *etc*.
- **Cross-validation:** Check the model's ability to generalize well by subjecting it to tests on different subsets of data.

Optimizing model parameters is essential to achieving accurate and reliable energy predictions in hybrid solar-wind systems [58 - 60].

- Model Evaluation and Performance Metrics

After training, models are evaluated using various performance metrics to assess accuracy and reliability:

- **Forecasting models:**
- Mean Absolute, Percentage Error (MAPE)
- Coefficient of Determination (R^2)
- Root Mean Square Error (RMSE)
- **Optimization models:**
- Energy savings percentage
- Reduction in dependency on fossil fuel backup
- Computational efficiency (time taken to optimize system performance)

AI-based forecasting models have improved hybrid system reliability by up to 30% in real-world applications [55]. Fig. (**5**) illustrates the workflow of machine learning and deep learning techniques applied within a hybrid system framework. The process includes data collection, preprocessing, feature extraction, model training, validation, and deployment for optimized system performance.

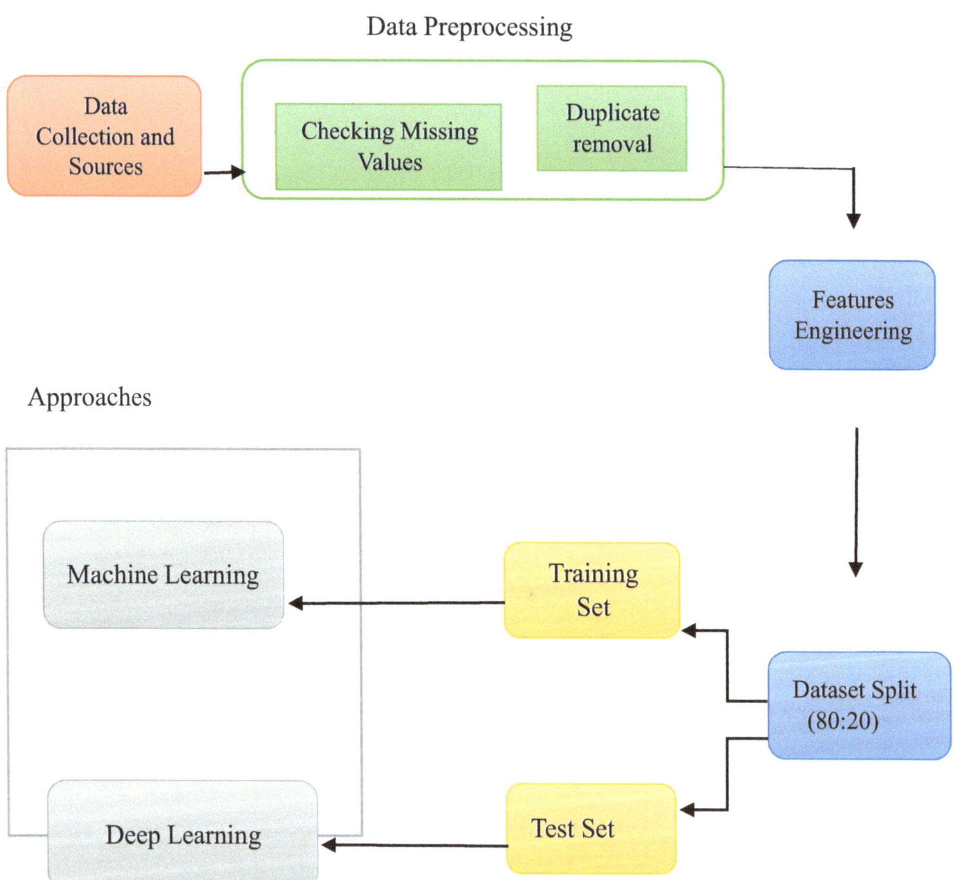

Fig. (5). Process involved in ML and DL in hybrid system.

APPROACHES TO ACHIEVE HYBRID SOLAR-WIND SYSTEM POWERED BY AI

AI-Based Energy Forecasting

The fundamental way to ensure the stability and efficiency of hybrid energy systems is the correct projection of solar and wind energy generation. Forecasting power generation using both planned and historical meteorological data has been accomplished with several ML and DL techniques. Artificial neural networks, support vector machines, and models such as random forests are very commonly used for predicting the generation of solar and wind energy with good accuracy [34].

Machine Learning Approaches for Forecasting

- **Artificial Neural Networks (ANNs):** Because they have the ability to model very complex and non-linear relationships between weather variables and the production of energy, ANNs are now widely used in hybrid energy forecasting. ANN-based models have shown promising results in predicting energy output by learning from past trends and adjusting dynamically to new data [35].
- **Support Vector Machines (SVMs):** These are effective for regression-based energy forecasting and are particularly useful when dealing with limited datasets. Support vector regression (SVR) is successfully applied to hybrid solar-wind power systems for precise energy output predictions [37].
- **Random Forests (RFs):** When combining different decision trees, they create random forests, which enhance the accuracy achieved in this method. Moreover, they are much more resistant to overfitting as well as noise in any meteorological data. Random forest algorithms have demonstrated high forecasting accuracy by leveraging ensemble learning techniques [39].
- **Extreme Gradient Boosting (XGBoost):** Similar to that of time-series analysis, bias and variance have to be reduced to increase accuracy in prediction, and this is where XGBoost comes in handy. XGBoost-based models have been effectively used for renewable energy forecasting, enhancing the reliability of hybrid solar-wind systems [53]. Fig. (**6**) presents various machine learning models commonly used for energy forecasting, including regression, decision trees, support vector machines, and neural networks. These models enable improved accuracy by capturing nonlinear relationships in energy data.

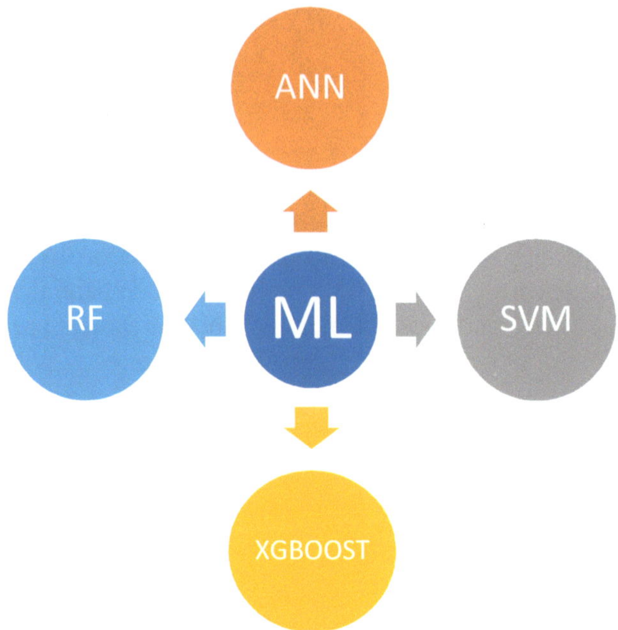

Fig. (6). Machine learning (ml) models for energy forecasting.

Deep Learning Approaches to Forecasting

- **Long Short-Term Memory Networks (LSTM):** LSTM is a form of Recurrent Neural Network (RNN) utilized for time series forecasting; hybrid systems comprising actual and reactive components can be time-variant. LSTM-based forecasting models have significantly improved hybrid solar-wind energy predictions by capturing long-term dependencies in weather datasets [41].
- **Convolutional Neural Networks (CNNs):** CNNs are pulling spatial features from satellite images and meteorological maps, which can help in the creation of a forecasting. CNN-based hybrid forecasting models integrate spatial-temporal data, achieving higher precision in solar and wind power generation predictions [42].
- **Transformer Networks**: Energy forecasting has made use of transformer models, which are similar to those used in natural language processing. Transformer-based models have demonstrated superior performance in hybrid solar-wind forecasting by effectively capturing temporal dependencies in meteorological data [56].
- **Hybrid Deep Learning Models:** The research uses combining different deep learning (DL) techniques, such as CNN-LSTM, to obtain accurate results. Hybrid AI models integrating CNN and LSTM architectures have outperformed standalone forecasting models in renewable energy applications [44].

- **Deep Learning in Offshore Wind-Wave Systems:** Deep learning models have also been applied to offshore hybrid energy systems, improving efficiency and power prediction. Deep Learning based models have great potential in offshore hybrid wind-wave energy systems, enhancing predictive accuracy for optimal energy generation [52].

AI-BASED OPTIMIZATION TECHNIQUES

Thus, these hybrid energy systems generally revolve around economical allocation-distribution of resources and power among other parameters. Numerous ML-DL algorithms have been utilized to fully optimize energy storage, grid operation, and load balancing. The next demand level in the system can be predicted using the previous demand level. This associated demand level can vary with time. Hybrid energy system optimization refers to the efficient distribution and allocation of power as well as resources. Quite a few ML-DL algorithms enhance efficiency regarding energy storage, grid operation, and load balancing.

Hybrid energy system optimization generally deals with optimal allocation and distribution of resources and energy. Several ML and DL algorithms can potentially optimize the operations of energy storage, grid operation, and load balancing.

Machine Learning Optimization Algorithms:

- **Genetic Algorithms (GA):** GA optimizes the hybrid solar-wind system setup by simulating natural selection. Genetic algorithm-based optimization has successfully enhanced system efficiency by determining the optimal size of solar panels & wind turbines [45].
- **Particle Swarm Optimization (PSO):** The concept of PSO is for population-based algorithms where the parameters of a system arc iteratively tweaked to minimize energy losses and maximize efficiency. Particle swarm optimization has been widely implemented in hybrid energy management, leading to cost-effective and adaptive system control [46].

- **Bayesian Optimization:** The Bayesian inference tries to analyze values that optimally enhance the predictions of the energy consumption model, thereby ensuring more accurate forecasts. Bayesian optimization techniques have been successfully employed to enhance deep learning models for energy system forecasting [58].
- **Reinforcement Learning (RL):** RL is basically keeping in touch with its learning and changes in real-time concerning energy generation and consumption when the latter is in the input state. RL algorithms are always learning and adapting in real-time with respect to variations in energy generation

as well as demand. Deep reinforcement learning models have shown significant improvements in energy dispatch strategies, reducing power losses and enhancing grid reliability [47].

Deep Learning for Optimization

- **Deep Q-Networks (DQN):** To tolerate the energy trading and dynamic pricing strategies in hybrid systems. DQN models have enabled self-learning energy trading mechanisms, improving economic efficiency in renewable energy markets [48].
- **Autoencoders:** The system is designed to assist in the detection of faults and anomalies in hybrid solar-wind systems that can also facilitate the modernization of maintenance strategies. Discuss the analysis of faults and anomalies in hybrid systems of solar-wind energy. This analysis enables the enhancement of maintenance strategies. Deep autoencoder networks have been applied to identify system faults early, reducing downtime and maintenance costs [49].

AI-POWERED DECISION-MAKING IN HYBRID ENERGY SYSTEMS

The applications of AI techniques, machine learning, and deep learning algorithms are involved in making decisions concerning hybrid energy networks. Smart electric grid systems, characterized by Artificial Intelligence, can automatically adjust to revised dynamics in supply or demand with respect to generation resources. AI-driven decision-making frameworks enable real-time load balancing, ensuring efficient energy utilization in hybrid solar-wind microgrids [50].

- **Fuzzy Logic Systems:** Handle uncertainty and imprecise data in energy decision-making. Fuzzy logic-based decision models have been successfully integrated into hybrid systems, improving adaptability to unpredictable energy variations [51].
- **Self-Optimizing Grids:** AI-based models adjust grid parameters dynamically for improved efficiency. Self-optimizing grid systems leverage AI to minimize energy losses and enhance reliability in renewable energy networks [52]. Table **2** compares the performance and characteristics of ML and DL approaches applied to hybrid solar-wind energy systems. The comparison highlights differences in accuracy, computational complexity, data requirements, and adaptability to varying environmental conditions.

Table 2. Comparison of ML and DL approaches in hybrid solar-wind systems.

Aspect	Machine Learning Approaches	Deep Learning Approaches
Common Algorithms	Decision Trees, Random Forest, SVM, KNN, Gradient Boosting, XGBoost	ANN, CNN, RNN, LSTM, GRU, Transformer-based Models
Data Requirements	Requires structured and labelled data; works well with moderate datasets.	Needs large datasets; performs well with both structured and unstructured data.
Feature Engineering	Selection and extraction of features must be done manually.	Automatically extracts important features.
Training Time	Faster training time with small to medium datasets.	Computationally intensive, requires more training time
Computational Cost	Lower cost; can run on standard CPUs	High cost; requires GPUs/TPUs for efficient processing
Performance in Forecasting	Works well for short-term predictions	Excels in long-term time-series forecasting
Application in Hybrid Systems	Used for demand forecasting, weather prediction, and energy consumption optimization.	Used for fault detection, anomaly detection, and real-time adaptive control.
Suitability	Suitable for simpler, rule-based energy optimization models.	Best for advanced AI-driven hybrid energy management.
Scalability	More scalable with traditional optimization methods.	Scales well with large datasets but requires more resources.

ADVANTAGES OF HYBRID SOLAR-WIND SYSTEM POWERED BY AI OVER TRADITIONAL METHODS

Despite the old methodologies in terms of efficiency, adaptability, and predictive capacity, Artificial Intelligence would obviously stir development, and along with it, the management of hybrid solar-winds energy systems. Optimization through Artificial Intelligence falls into place in that it improves energy utilization, forecasting accuracy, and grid steadiness, all conditions unavailable in traditional approaches.

- **Enhanced Predictive Analytics and Forecasting:** Now, those models use Long Short-Term Memory (LSTM) and Transformer Networks in judging the historical weather patterns and the available meteorological data, and these predictions are more accurate than any rule-based or simple statistical means for the popular outputs of solar and wind power. AI forecasts will recognize certain climate variability and some unforeseen weather-change factors to better choose between power generation and storage management. AI-based forecasting models have improved hybrid system reliability by up to 30% in real-world applications [55].

- **Dynamic Optimization of Energy Dispatch:** The current methods of energy dispatch do employ rigid threshold values, which eventually induce some lag of energy wastage or shortages. AI-based optimization methods, like Genetic Algorithms (GA) and Particle Swarm Optimization (PSO), as well as Deep Reinforcement learning (DRL), set atop dynamic allocations in energy unit allocation and function in real time with loads and supplies. AI would hold that real-time balancing of loads to a more efficient distribution scheme between solar, wind, and storage units, and reduce the reliance on fossil backup. AI-powered optimization techniques dynamically adjust energy production based on real-time demand and environmental conditions, significantly improving efficiency [39].
- **Improved Grid Stability and Smart Energy Management:** AI-enabled self-healing grid management constantly identifies grid disturbances, predicts potential system faults, and prevents blackouts through early fault detection. Deep learning models are constantly used to monitor energy flows with CNN for anomaly detection to detect faults and enhance power quality in hybrid microgrids. Distancing their systems from hybrid systems that respond when real power fluctuations happen, AI-based systems now anticipate and respond to power fluctuations before they are actualized. Neural network-based control models enhance grid stability by predicting power fluctuations and adjusting energy flow accordingly [42].
- **Adaptive Learning and Continuous Improvement:** Unlike today, traditional hybrid energy systems indeed use a set of fixed, pre-defined rules. The hybrid energy systems thus have no ability to include future updates such as the varying weather conditions or loading condition change. AI models have thus continued adapting their configuration with the passage of time, based on real-time data, which implies they continue achieving improvement in efficiency with the passage of time. Reinforcement Learning: AI-driven controllers automatically adjust in the long term by self-optimizing using RL based on how well they are performing in real life.
- **Efficient Energy Storage Management:** AI Battery Management Systems forecast energy demand variation and adapt the charge-discharge cycles ahead of time to extend the lifespan of a battery. Contrary to the conventional rule-based ones, BMSs don't fail to reach their peak efficiency storage, therefore provoking recurrent deep discharges or over-writings. Predictive battery scheduling using AI will maximize the storage and utilization of energy, enhancing the reliability of the entire system. Machine learning-based battery management systems have enabled predictive maintenance and dynamic charge-discharge scheduling, extending battery lifespan and improving energy

availability [44].

- **Reduced Operational Costs and Maintenance:** Predictive maintenance seeks to forestall malfunctions from happening through AI monitoring for unusual behaviours ahead of systems being disrupted or entirely unusable. Compared to classic preventive approaches, which involve an equipment periodic downgrading with real-time diagnostics, predictive maintenance using AI cuts repairs by somewhere between 20% to 30%. AI-driven hybrid energy systems have led to an increase in efficiency by 25% and a 30% reduction in operational costs [36].
- **Scalability and Flexibility for Large-Scale Deployment:** Conventional hybrid energy models have limited scalability at the large scale. Highly scalable: AI algorithms offer seamless integration to grid-scale renewables projects themselves. AI offers hybrid energy management optimization in real-time across various locations to ensure balanced energy distribution across changing geographical conditions.
- **Regulatory Compliance and Smart Policy Adaptation:** AI energy management systems would process real-time regulatory change and adapt system parameters automatically to meet the energy policies. Unlike the manual method, AI-based frameworks are the ones that update energy models continuously based on changes in environmental regulations. Table **3** compares traditional hybrid energy systems with AI-powered counterparts across key operational and performance aspects.

Table 3. Comparison of AI-powered *vs.* traditional hybrid system.

Aspect	Traditional Methods	AI-Powered Hybrid Systems
Forecasting Accuracy	Limited, relies on static models.	High, dynamic and real-time AI-based predictions.
Energy Distribution	Rule-based, inefficient allocation.	Optimized using ML/DL for dynamic balancing.
Grid Stability	Reactive, handles fluctuations poorly.	Proactive, prevents failures using AI monitoring.
Adaptability	Fixed operational thresholds.	Self-learning adjusts to changing conditions.
Battery Management	Static charge-discharge cycles.	Predictive optimization extends battery life.
Operational Costs	High, requires manual maintenance.	Lower due to AI-driven predictive maintenance.
Scalability	Difficult to scale efficiently.	AI enables seamless large-scale expansion.
Regulatory Compliance	Manual updates required	AI adapts automatically to new policies.

ENVIRONMENTAL AND ECONOMIC IMPACTS OF AI-POWERED HYBRID SOLAR-WIND SYSTEMS

AI optimization in hybrid energy systems tends to make ecosystems much more sustainable in terms of decreasing their environmental footprint, improving energy efficiency, and yielding economic returns. AI-powered environmental impact assessments have successfully minimized land-use conflicts, optimizing energy farm locations with reduced ecological disruption [46].

- **Reducing CO_2 emissions and relying on fossil fuels in general:** Energy forecasting and management of storage of AI minimizes reliance on fossil-fue--based backup systems through the optimization of the energy generation scheme. From reports, there is strong evidence that carbon emissions using AI-powered hybrid systems are reduced by no more than 30%. The argument put forth is on improved energy efficiency and lower curtailment losses. Optimized AI-driven hybrid energy systems have shown a measurable reduction in carbon footprint by improving renewable energy utilization and minimizing fossil fuel dependency [48].
- **Enhancing Energy Efficiency and Production**: Machine-learning algorithms predict the supply of solar and wind energies for maximum efficient use of resources. AI dispatch optimization will enhance efficiency by about 5 to 15 percent compared to what we have now on annual energy output. AI-driven forecasting and adaptive energy dispatch have significantly increased hybrid energy efficiency, reducing unpredictability in renewable power generation [50].
- **Reduced Construction and Operating Costs:** AI in site selection helps put infrastructure construction at a low cost, thus minimizing initial capital. Predictive Maintenance by Artificial Intelligence enables the early detection of potential failure events, thereby reducing downtimes of operations and having a possible restriction on operational costs by 20-30%. AI-enabled predictive maintenance strategies have reduced repair costs and extended equipment lifespan, ensuring greater cost-efficiency in hybrid renewable systems [52].
- **Elevating Financial Feasibility and Investment Margins:** Energy foresight with Artificial Intelligence offers near-perfect energy yield predictions, giving hybrid systems more attractiveness to potential investors willing to finance the program. Research has evidenced a subsequent 30% increase in ROI from efficient allocation of resources. Investment in AI-optimized hybrid energy projects has resulted in a significant improvement in financial viability, with higher ROI and reduced economic risks [54].
- **Global Flexibility to Climate and Resilience to Weather Adversities**: AI blends historical weather data with real-time climate information that helps forecast changes in the environment over an extremely long time. Hence, hybrid

systems can adapt to seasonal fluctuations and extreme weather conditions so as to affect long-term resilience and sustainability. AI-driven hybrid energy systems demonstrate superior resilience to climate variability by leveraging adaptive control mechanisms for energy generation [56].

APPLICATIONS OF AI IN HYBRID SOLAR-WIND ENERGY: OPTIMIZATION, EFFICIENCY, AND SUSTAINABILITY

Artificial Intelligence presents a breakthrough approach to hybrid solar systems in terms of enhanced optimization, efficiency optimization technologies, and effective management procedures. It allows a smart energy management based on the algorithms functioning around detection and processing a large set of data, including weather, demand, and operational. Unlike static rules of the traditional models, which make them find hybrid energy systems more attractive in terms of economic and environmental values based on cost-effectiveness, AI is up to date with ad hoc decisions.

- **AI for Energy Forecasting and Demand Prediction**

Proper energy prediction is required to gain maximum efficiency and profits from renewable energy in its erratic nature. These AI models, such as those of Long Short-Term Memory networks (LSTM), Support Vector Machines (SVMs), or Random Forest, would likely require historical data regarding the previous weather, solar irradiance, and wind-speed patterns- all of which would be inputs for it to predict energy output with absolute precision. In an analysis of what AI-based forecasting means, it means less wastage on the energy production side, but a great proportion of those who really need the power. A Transformer-based approach will improve predictive accuracy since this model absorbs the risk of fluctuating patterns of less and more solar irradiance or wind speed while relying on helpful battery storage systems and minimizing or eliminating the need for a backup supply system for oil. Studies indicate that AI-based forecasting has led to a 25% reduction in energy wastage while improving grid reliability in hybrid solar-wind systems [46].

- **AI-Based Grid Optimization and Load Balancing**

Not being able to manage the irregularity of renewable energy sources, because solar and wind are the most crucial problems in grid management. It is an intrinsic inefficiency that underpins this feature. The set time-by-time instability inherent in the grid itself is further conditioned. Optimizing the power distribution system with a much higher level of stability and efficiency is achieved through AI-associated reduction of energy flow with increasing energy demand. Load balancing can be made 'smart' through predictive analytics, such that input parameters will analyze the consumption pattern through Artificial Intelligence and then discharge the required power. This should go a long way in reducing possible changes in voltage profile, frequency imbalances, and high rates of transmission losses. Research has shown that machine learning models applied to grid optimization have enhanced energy efficiency, minimized congestion, and improved overall grid performance in hybrid renewable systems [48].

- **Intelligent Energy Dispatch and Storage Optimization**

Dispatch management in energy is an important requirement for tapping the current capacity of hybrid solar-wind power systems. Energy allocation considering solar panels, wind turbines, and battery systems can be taken on a distribution factor using AI-specific evolutionary algorithms such as Genetic Algorithm (GA) and Particle Swarm Optimization (PSO). Energy assignment in a timely manner is facilitated by models that are updated each and every time with current weather forecasts and real-time power requirements to prevent shortages and excess generation during grid dispatch. Battery units are charged and discharged by case BMS systems that minimize the wastage of energy and at the same time prolong the lifetime of batteries. Currently, an AI predictive scheduling tool is applied to ensure that the efficiency of battery storage systems is preserved. Studies have found that AI-based energy storage optimization can extend battery lifespan by 40% while reducing unnecessary charging cycles, leading to improved system performance and reliability [50].

- **Predictive Maintenance & Fault Detection**

To the best of its advantage, Artificial Intelligence shaped the maintenance philosophy in hybrid-energy systems, guaranteeing a decrease in failure scenarios and hence a decrease in operational costs, as expectedly. Using extremely effective AI-based anomaly detection systems, energy infrastructures can be Facebook, LinkedIn! Safely, all activities monitor the energy resources and respond at the point of emergence of the irregularities that may later become catastrophic faults. Damage predictions use deep learning models that process sensor data from solar panels and wind turbines. With real-time input for preparation for preventive maintenance, drone-based inspection systems may monitor turbine blades and solar panels for their efficiencies. AI-enabled fault detection has demonstrated a reduction in maintenance costs by up to 30%, significantly increasing the overall reliability of hybrid energy systems [52].

- **AI Powers Sustainability and Adaptability to the Environment**

Where AI is concerned, it is fully up to enhancing the sustainability of hybrid-solar wind systems with the impact assessment of the environment, including studies of site selection, since AI completely enhances the integration of such models. Such a rotation in site selection models can benefit from the increasing degrees of predictability and the direct knowledge of where to install for maximum energy production, least environmental disturbance through site selection AI, which comprises algorithms that evaluate terrain, weather patterns, and land-use constraints in selecting out a site that has the least environmental footprint. Finally, through hybridization, AI allows use in terms of future climatic variations that would be forecasted as climate adaptation mechanisms to portray the projection. AI-optimized hybrid energy systems have been shown to improve sustainability metrics by increasing renewable energy utilization, reducing land-use conflicts, and enhancing long-term energy security [54]. Fig. (7) highlights the key advantages of Artificial Intelligence, including enhanced predictive accuracy, automation of complex tasks, real-time decision-making, improved efficiency, and the ability to process large and diverse datasets.

CHALLENGES AND FUTURE DIRECTIONS IN AI-POWERED HYBRID SOLAR-WIND SYSTEMS

Though the promise of Artificial Intelligence applications in boosting the operational efficiency of a hybrid energy system appears bright, a series of myriad challenges need resolution before most of the large-scale implementations. Such challenges run the gamut, from those related to data, through limitations in computational requirements, to those concerning security. Some of the more profound hindrances include:

- **Data Availability and Quality Issues**

AI models for the hybrid solar-wind system rely on high-quality weather, geographical, and operational data; inconsistencies in the methods of data collection, with several values missing or accessible to real-time meteorological datasets for only a limited time, render the AI prediction and optimization less reliable [45].

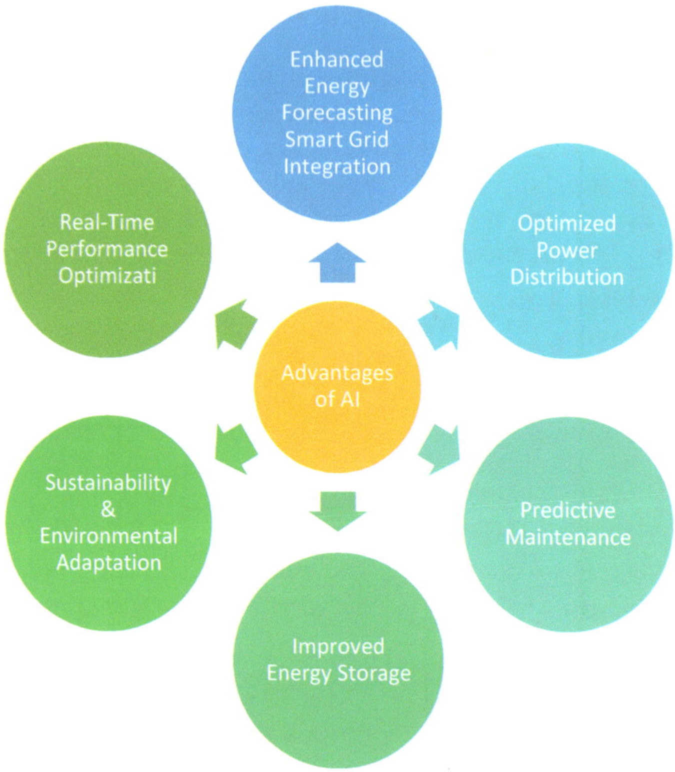

Fig. (7). Advantages of Artificial Intelligence.

• Uncertainty in Renewable Energy Forecasting

Forecasting power from solar and wind sources is actually pure gold for energy management. However, the stochastic behaviour of weather conditions, seasonal variation, and extreme climatic events constitutes a great hindrance to forecasting. Consequently, in extreme cases, uncertainty in energy prices significantly disturbs the energy dispatch planning and grid stability [46].

• High Computational Requirements

Extensive computational resources will be required for the implementation of large-scale spatiotemporal datasets by complex AI models utilizing numerous variations of the deep-learning paradigm for optimization. Such high computational demand forms one of the most critical hurdles, particularly in the case of off-grid or remote energy systems where such resources are limited [47].

- **Integration with Existing Infrastructure**

Most of the existing hybrid energy systems are based on conventional control schemes. Their integration with automation and AI-driven predictive analytics will thus require huge refurbishments, equally increasing the installation cost. The degree of incompatibility with legacy systems would form a further hindrance to its adoption [48].

- **Cybersecurity Risks**

An AI-based smart grid system relies on interconnecting an IoT, cloud, and wireless communication network, all of which remain susceptible to cyber-attacks. Unauthorized persons infringing on the AI-based energy management systems may disrupt the power supply and compromise sensitive energy data [49].

As some solutions for these challenges would aim to increase the efficiency of hybrid solar-wind systems, further AI advances would consider such approaches:

- **Development of Standardized and Open-Source Datasets**

Future research should develop standardized databases for renewable energy resources that grant public access. Improvements in satellite imaging can offer direct enhancement of data availability and accuracy in conjunction with IoT environmental sensing and blockchain data-sharing technologies.

- **Hybrid AI Models for Improved Forecasting**

Hybrid AI modelling, which uses the physics of weather patterns, will be a better approach for renewable energy prediction. Hybrid models that couple traditional methods of forecasting with machine learning approaches will improve the short- and long-term forecasting reliability.

- **Cloud and Edge Computing for Real-Time Optimization**

Hosting AI modules on cloud platforms and edge computing devices will diminish latency and computational burden. If the AI-driven optimization emanates from data context produced in situ at the generation site, it can be validated in real-time, avoiding excessive stress for further hardware.

• Explainable AI (XAI) for Enhanced Transparency

Some of the AI algorithms will have their decision-making processes elucidated with the application of XAI, which will empower energy stakeholders to better comprehend how the recommendations from AI are given. This will also aid in sustaining legitimacy and complying with regulatory standards in hybrid energy systems.

• AI-Driven Cybersecurity Enhancements

Future enhanced AI measures are expected to provide cyber protection for hybrid solar-wind systems by timely detecting anomalies of cyber threats for secure data transmissions. AI intrusion-detection systems, coupled with blockchain security frameworks, can boost the resilience of energy infrastructure against cyber threats.

Doing away with the constraints subsequently mentioned will take these exploratory directions of the future and make the AI-enabled hybrid solar-wind systems feasible, secure, and efficient, injecting vigour into the global energy transition towards sustainable energy alternatives.

CONCLUSION

They will help in redefining the ambient chemistry in the IoT flavour. Decreased fuel dependency, decreased carbon sinks, and surely improved efficiencies of soil consumption were due to these transformations. The return on investment and risk would place AI as a major facilitator toward an energy transition for the world. Unmasking AI with its capacities over renewable energy will, however, have an extra burden over cyber threats. Development for the future should see the coupling of AI into hybrid forecasting models with physics-based techniques to develop energy prediction accuracies. Furthermore, with edge computing, real-time energy optimization will become possible even with lower computing costs. Wider acceptance and regulation compliance could also facilitate XAI, which will be required to ensure transparency and trust regarding AI decisions. Hybrid solar-wind energy systems influenced by AI are likely to bring about a transformation in meeting world energy demand with the most sustainable, efficient, and cost-effective alternatives to conventional energy sources. In this way, research and technological advancements would help minimize barriers to the widespread adoption of AI-powered renewable energy systems, promoting a sustainable and resilient future.

REFERENCES

[1] N.H. Afgan, D.A. Gobaisi, M.G. Carvalho, and M. Cumo, "Sustainable energy development", *Renew. Sustain. Energy Rev.,* vol. 2, no. 3, pp. 235-286, 1998.

[http://dx.doi.org/10.1016/S1364-0321(98)00002-1]

[2] D. Raimi, E. Campbell, R. Newell, B. Prest, S. Villanueva, and J. Wingenroth, *Global energy outlook 2022: turning points and tension in the energy transition.* Resources for the Future: Washington, DC, USA, 2022, pp. 1723-1742.

[3] I. U. Haq, I. U. Hassan, and H. A. Shah, "Machine Learning Techniques for Result Prediction of One Day International (ODI) Cricket Match", In: *Proceedings of the 2023 IEEE 8th International Conference for Convergence in Technology (I2CT)*, 2023, pp. 1-5. [http://dx.doi.org/10.1109/I2CT57861.2023.10126241]

[4] B.K. Sovacool, "How long will it take? Conceptualizing the temporal dynamics of energy transitions", *Energy Res. Soc. Sci.,* vol. 13, pp. 202-215, 2016. [http://dx.doi.org/10.1016/j.erss.2015.12.020]

[5] M.Z. Jacobson, M.A. Delucchi, M.A. Cameron, and B.A. Frew, "Low-cost solution to the grid reliability problem with 100% penetration of intermittent wind, water, and solar for all purposes", *Proc. Natl. Acad. Sci. USA,* vol. 112, no. 49, pp. 15060-15065, 2015. [http://dx.doi.org/10.1073/pnas.1510028112] [PMID: 26598655]

[6] S. J. Russell, and P. Norvig, "Artificial Intelligence: A Modern Approach", In: *Pearson* 3rd. Upper Saddle River, NJ, USA: Pearson, 2016.

[7] F. Li, S. Chen, X. Zhang, W. Huang, and G. Ma, "Short-term optimal coordinated operation of a wind-solar-hydro hybrid system based on deep learning and double-layer nesting solution algorithm", *Research Square,* 2022. [http://dx.doi.org/10.21203/rs.3.rs-1502713/v1]

[8] K. Salah, M.H.U. Rehman, N. Nizamuddin, and A. Al-Fuqaha, "Blockchain for AI: Review and open research challenges", *IEEE Access,* vol. 7, pp. 10127-10149, 2019. [http://dx.doi.org/10.1109/ACCESS.2018.2890507]

[9] Y. Zhu, M. Wang, X. Yin, J. Zhang, E. Meijering, and J. Hu, "Deep learning in diverse intelligent sensor based systems", *Sensors (Basel),* vol. 23, no. 1, p. 62, 2022. [http://dx.doi.org/10.3390/s23010062] [PMID: 36616657]

[10] I. U. Haq, A. H. Rather, and G. Kaur, "A Comparative Analysis of Machine Learning Algorithms for the Early Prediction of Cardiovascular Disease", In: *Proceedings of the 2023 2nd International Conference on Edge Computing and Applications (ICECAA)*, 2023, pp. 987-993. [http://dx.doi.org/10.1109/ICECAA58104.2023.10212320]

[11] Y. LeCun, Y. Bengio, and G. Hinton, "Deep learning", *Nature,* vol. 521, no. 7553, pp. 436-444, 2015. [http://dx.doi.org/10.1038/nature14539]

[12] D. Jurafsky, and J. H. Martin, "Speech and Language Processing: An Introduction to Natural Language Processing, Computational Linguistics, and Speech Recognition with Language Models", In: *Pearson* 3rd ed. draft. Stanford, CA, USA: Pearson, 2025.

[13] R. Szeliski, "Computer Vision: Algorithms and Applications", 2nd ed. Washington, DC, USA: Springer Nature, 2022.

[14] B. Siciliano and O. Khatib (Eds.), Springer *Handbook of Robotics*, Berlin, Germany: Springer-Verlag, 2008.

[15] I. Sapre, "The Role of Artificial Intelligence in Enhancing Renewable Energy Efficiency: A Case Study on Solar and Wind Energy Optimization. Intersect: The Stanford Journal of Science", *Technol. Soc.,* vol. 18, no. 1, 2024.

[16] S.K. Mulpuri, B. Sah, and P. Kumar, "An intelligent battery management system (BMS) with end-edge-cloud connectivity – a perspective", *Sustain. Energy Fuels,* vol. 9, no. 5, pp. 1142-1159, 2025. [http://dx.doi.org/10.1039/D4SE01238K]

[17] S. E. Ali, O. G. Ndubuisi, C. A. R. Obiorah, U. T. Aku, O. Nesiama, C. O. Agbakhamen, and O. P.

Okoro, "Optimizing Energy and Resource Utilization for a Sustainable Future", *Int. J. Innov. Math. Stat. Energy Policies,* vol. 13, no. 1, pp. 72-79, 2025.

[18] A. Masood, U. Ahmed, S. Z. Hassan, A. R. Khan, and A. Mahmood, "Economic Value Creation of Artificial Intelligence in Supporting Variable Renewable Energy Resource Integration to Power Systems: A Systematic Review", *Sustainability*. vol. 17, no. 6, Art. no. 2599, Mar. 2025. [http://dx.doi.org/10.3390/su17062599]

[19] J. Nilimaa, and V. Zhaka, "An overview of smart materials and technologies for concrete construction in cold weather", *Eng,* vol. 4, no. 2, pp. 1550-1580, 2023. [http://dx.doi.org/10.3390/eng4020089]

[20] A. H. Rather, and I. U. Haq, "Intelligent Framework for Early Prediction of Diabetic Retinopathy: A Deep Learning Approach", In: *Proceedings of International Conference on Paradigms of Communication, Computing and Data Analytics (PCCDA 2023)*. Springer: Singapore, 2023, pp. 107-117. [http://dx.doi.org/10.1007/978-981-99-4626-6_32]

[21] I. R. Saleh, B. Rafiei, K. Gharali, and B. Sajadi, "Economic feasibility of solar and wind energy harvesting in Karbala, Iraq", *Ain Shams Eng. J.*. vol. 17, no. 1, Art. no. 103946, Jan. 2026. [http://dx.doi.org/10.1016/j.asej.2025.103946]

[22] A. Javaid, M. Sajid, E. Uddin, A. Waqas, and Y. Ayaz, "Sustainable urban energy solutions: Forecasting energy production for hybrid solar-wind systems", *Energy Convers. Manage.,* vol. 302, p. 118120, 2024. [http://dx.doi.org/10.1016/j.enconman.2024.118120]

[23] M. S. H. Lipu, M. S. Miah, M. A. Hannan, A. Hussain, M. R. Sarker, A. Ayob, M. H. M. Saad, and M. S. Mahmud, "Artificial intelligence based hybrid forecasting approaches for wind power generation: Progress, challenges and prospects", *IEEE Access,* vol. 9, pp. 102460-102489, 2021. [http://dx.doi.org/10.1109/ACCESS.2021.3097102]

[24] M. D. A. Al-falahi, S. D. G. Jayasinghe, and H. Enshaei, "A review on recent size optimization methodologies for standalone solar and wind hybrid renewable energy system", *Energy Convers. Manage.,* vol. 143, pp. 252-274, 2017. [http://dx.doi.org/10.1016/j.enconman.2017.04.019]

[25] I. Ul Haq, "Diagnosis of Neurological Disease Using Bioinspired Algorithms", In: *Bio-Inspired Optimization for Medical Data Mining,* S. Srivastava, A. Anand, A. Kumar, B. Saini, P. S. Rathore, Eds., Wiley-IEEE Press: Hoboken, NJ, USA, 2024, pp. 245-270. [http://dx.doi.org/10.1002/9781394213894.ch11]

[26] A.U. Chávez-Ramírez, V. Vallejo-Becerra, J.C. Cruz, R. Ornelas, G. Orozco, R. Muñoz-Guerrero, and L.G. Arriaga, "A hybrid power plant (Solar–Wind–Hydrogen) model based in artificial intelligence for a remote-housing application in Mexico", *Int. J. Hydrogen Energy,* vol. 38, no. 6, pp. 2641-2655, 2013. [http://dx.doi.org/10.1016/j.ijhydene.2012.11.140]

[27] A. Sabo, B. Kolapo, T. Odoh, M. Dyari, N. Abdul Wahab, and V. Veerasamy, "Solar, wind and their hybridization integration for multi-machine power system oscillation controller's optimization: a review", *Energies,* vol. 16, no. 1, p. 24, 2022. [http://dx.doi.org/10.3390/en16010024]

[28] B. Oussmou, S. Sigue, and S. Abderafi, "Review of green hydrogen production technologies to choose the optimal process of electrolysis–renewable energy", *Renewable and Sustainable Energy Reviews,* vol. 225, p. 116205, 2026.

[29] S. Singh, M. Singh, and S.C. Kaushik, "A review on optimization techniques for sizing of solar-wind hybrid energy systems", *Int. J. Green Energy,* vol. 13, no. 15, pp. 1564-1578, 2016. [http://dx.doi.org/10.1080/15435075.2016.1207079]

[30] Khosravi, A. (2018). Prediction of solar/wind energy in a hybrid renewable energy system using

artificial intelligence.

[31] A. Maleki, and A. Askarzadeh, "Comparative study of artificial intelligence techniques for sizing of a hydrogen-based stand-alone photovoltaic/wind hybrid system", *Int. J. Hydrog. Energy,* vol. 39, no. 19, pp. 9973-9984, 2014. [http://dx.doi.org/10.1016/j.ijhydene.2014.04.147]

[32] P. S. Sikder, and N. Pal, "Modeling of an intelligent battery controller for standalone solar-wind hybrid distributed generation system", *J. King Saud Univ. Eng. Sci.,* vol. 32, no. 6, pp. 368-377, 2020. [http://dx.doi.org/10.1016/j.jksues.2019.04.004]

[33] M.A. Elhadidy, "Performance evaluation of hybrid (wind/solar/diesel) power systems", *Renew. Energy,* vol. 26, no. 3, pp. 401-413, 2002. [http://dx.doi.org/10.1016/S0960-1481(01)00139-2]

[34] T. P. Fudge, and S. S. Ferebee, "Solar and wind energy in urban transformation: Opportunities, challenges, and cybersecurity", *Navigating Cybersecurity and Privacy in the Evolution of Smart Urban Ecosystems.,* pp. 271-310, 2026.

[35] V. Simankov, P. Buchatskiy, A. Kazak, S. Teploukhov, S. Onishchenko, K. Kuzmin, and P. Chetyrbok, "A solar and wind energy evaluation methodology using artificial intelligence technologies", *Energies,* vol. 17, no. 2, p. 416, 2024. [http://dx.doi.org/10.3390/en17020416]

[36] M. Li, Z. Xu, S. Li, Y. Kikuchi, Y. Dong, K. C. Gryllias, and J. Carroll, "Health prognostics and maintenance decision-making for wind energy: A comprehensive overview", *Renew. Sustain. Energy Rev.,* vol. 226, p. 116269, 2026.

[37] M. S. Mauludin, M. Khairudin, R. Asnawi, W. A. Mustafa, and S. F. Toha, "The Advancement of Artificial Intelligence's Application in Hybrid Solar and Wind Power Plant Optimization: A Study of the Literature", *J. Adv. Res. Appl. Sci. Eng. Technol,* vol. 50, no. 2, pp. 279-293, 2024. [http://dx.doi.org/10.37934/araset.50.2.279293]

[38] I. U. Haq, and A. H. Rather, "Generative AI-driven frameworks for intelligent wind energy forecasting and optimization", In: *Enhancing Renewable Energy Systems With Generative AI*, J. R. Juárez Ramírez, A. Kumar, P. S. Rathore, Eds., Hershey, PA, USA: IGI Global, 2026, pp. 403–426. [http://dx.doi.org/10.4018/979-8-3693-5182-6.ch017]

[39] A. Sabo, B. Kolapo, T. Odoh, M. Dyari, N. Abdul Wahab, and V. Veerasamy, "Solar, wind and their hybridization integration for multi-machine power system oscillation controllers optimization: a review", *Energies,* vol. 16, no. 1, p. 24, 2022. [http://dx.doi.org/10.3390/en16010024]

[40] A. Singh, I. U. Haq, N. Jain, R. Kumar, M. A. Rather, and F. S. Masoodi, "Hybrid quantum–classical learning architectures for 6G/7G edge", *Deep Quantum Neural Netw.: AI 6G/7G Commun. Syst.,* vol. 10, no. 10, p. 211, 2026.

[41] I. U. Haq, A. Kumar, and P. S. Rathore, "Machine learning approaches for wind power forecasting: A comprehensive review", *Discover Applied Sciences.,* vol. 7, no. 10, p. 1139, 2025.

[42] N. Sharma, M. Sharma, A. Singhal, R. Vyas, H. Malik, A. Afthanorhan, and M. A. Hossaini, "Recent trends in EEG-based motor imagery signal analysis and recognition: a comprehensive review", *IEEE Access,* vol. 11, pp. 80518-80542, 2023. [http://dx.doi.org/10.1109/ACCESS.2023.3297314]

[43] R. Singh, S.A. Memon, R. Shaikh, and D.S. Upadhyay, "A review of artificial intelligence applied for the solution of issues in the extensive adaption of solar and wind energy", *Int. J. Ambient Energy,* vol. 43, no. 1, pp. 7419-7436, 2022. [http://dx.doi.org/10.1080/01430750.2022.2063384]

[44] N.C. Ohalete, A.O. Aderibigbe, E.C. Ani, P.E. Ohenhen, D.O. Daraojimba, and B.A. Odulaja, "AI-driven solutions in renewable energy: A review of data science applications in solar and wind energy

optimization", *World J. Adv. Res. Rev*, vol. 20, no. 3, pp. 401-417, 2023. [http://dx.doi.org/10.30574/wjarr.2023.20.3.2433]

[45] T. Ilakkia, and G. Vijayagowri, "Hybrid PV/wind system for reduction of harmonics using artificial intelligence technique", In: *Proceedings of the 2012 International Conference on Advances in Engineering, Science and Management (ICAESM-2012)*, Nagapattinam. India, Mar. 16–17, 2012, pp. 303-308.

[46] A. Maleki, and A. Askarzadeh, "Comparative study of artificial intelligence techniques for sizing of a hydrogen-based stand-alone photovoltaic/wind hybrid system", *Int. J. Hydrog. Energy,* vol. 39, no. 19, pp. 9973-9984, 2014. [http://dx.doi.org/10.1016/j.ijhydene.2014.04.147]

[47] L. Abualigah, R.A. Zitar, K.H. Almotairi, A.M. Hussein, M. Abd Elaziz, M.R. Nikoo, and A.H. Gandomi, "Wind, solar, and photovoltaic renewable energy systems with and without energy storage optimization: A survey of advanced machine learning and deep learning techniques", *Energies,* vol. 15, no. 2, p. 578, 2022. [http://dx.doi.org/10.3390/en15020578]

[48] S. Siddula, G. K. Prashanth, P. Nandankar, R. Subbiah, S. M. Wabaidur, E. A. Al-Ammar, M. H. Siddique, and S. Thanappan, "Optimal Placement of Hybrid Wind-Solar System Using Deep Learning Model", *Int. J. Photoenergy,* vol. 2022, no. 1, pp. 1-7, 2022. [http://dx.doi.org/10.1155/2022/2881603]

[49] S. Shamshirband, T. Rabczuk, and K.W. Chau, "A survey of deep learning techniques: application in wind and solar energy resources", *IEEE Access,* vol. 7, pp. 164650-164666, 2019. [http://dx.doi.org/10.1109/ACCESS.2019.2951750]

[50] M. Abubakar, Y. Che, A. Zafar, M. A. Al-Khasawneh, and M. S. Bhutta, "Optimization of solar and wind power plants production through a parallel fusion approach with modified hybrid machine and deep learning models", *Intell. Data Anal.,* 2025. [http://dx.doi.org/10.1177/1088467X241312592]

[51] V. Upendran, M. C. M. Cheung, S. Hanasoge, and G. Krishnamurthi, "Solar wind prediction using deep learning", *Space Weather,* vol. 18, no. 9, p. e2020SW002478, 2020. [http://dx.doi.org/10.1029/2020SW002478]

[52] M. Dehghan Manshadi, M. Mousavi, M. Soltani, A. Mosavi, and L. Kovacs, "Deep learning for modeling an offshore hybrid wind–wave energy system", *Energies,* vol. 15, no. 24, p. 9484, 2022. [http://dx.doi.org/10.3390/en15249484]

[53] Y. Amoura, S. Torres, J. Lima, and A.I. Pereira, "Hybrid optimisation and machine learning models for wind and solar data prediction", *Int. J. Hybrid Intell. Syst.,* vol. 19, no. 1,2, pp. 45-60, 2023. [http://dx.doi.org/10.3233/HIS-230004]

[54] A. D. A. Bin Abu Sofian, H. R. Lim, H. Siti Halimatul Munawaroh, Z. Ma, K. W. Chew, and P. L. Show, "Machine learning and the renewable energy revolution: Exploring solar and wind energy solutions for a sustainable future including innovations in energy storage", *Sustain. Dev. (Bradford),* vol. 32, no. 4, pp. 3953-3978, 2024. [http://dx.doi.org/10.1002/sd.2885]

[55] A. Bhansali, N. Narasimhulu, R. Pérez de Prado, P.B. Divakarachari, and D.L. Narayan, "A review on sustainable energy sources using machine learning and deep learning models", *Energies,* vol. 16, no. 17, p. 6236, 2023. [http://dx.doi.org/10.3390/en16176236]

[56] D. A. Roberts, H. Karimabadi, T. Sipes, Y. K. Ko, and S. Lepri, "Objectively determining states of the solar wind using machine learning", *Astrophys. J.,* vol. 889, no. 2, p. 153, 2020. [http://dx.doi.org/10.3847/1538-4357/ab5a7a]

[57] T. Alazemi, M. Darwish, and M. Radi, Renewable energy sources integration *via* machine learning modelling: A systematic literature review, *Heliyon,* vol. 10, no. 4, p. e26088, 2024.

[http://dx.doi.org/10.1016/j.heliyon.2024.e26088] [PMID: 38404865]

[58] M.A. Rushdi, S. Yoshida, K. Watanabe, Y. Ohya, and A. Ismaiel, "Deep learning approaches for power prediction in wind–solar tower systems", *Energies,* vol. 17, no. 15, p. 3630, 2024. [http://dx.doi.org/10.3390/en17153630]

[59] H. Raju, and S. Das, "Cnn-based deep learning model for solar wind forecasting", *Sol. Phys.,* vol. 296, no. 9, p. 134, 2021. [http://dx.doi.org/10.1007/s11207-021-01874-6]

[60] I. Shafi, H. Khan, M.S. Farooq, I.T. Diez, Y. Miró, J.C. Galán, and I. Ashraf, "An artificial neural network-based approach for real-time hybrid wind–solar resource assessment and power estimation", *Energies,* vol. 16, no. 10, p. 4171, 2023. [http://dx.doi.org/10.3390/en16104171]

CHAPTER 9

Forecasting the Power Generation of Wind Turbines through Advanced Artificial Intelligence Techniques

Manik Rakhra[1], **Tiyas Sarkar**[1,*] and **Vikas Verma**[1]

[1] *School of Computer Science and Engineering, Lovely Professional University, Phagwara, Punjab, India*

Abstract: This analysis concerns the evaluation and modelling of wind turbine energy output using sophisticated Artificial Intelligence (AI) techniques. These include Machine Learning (ML), which makes use of polynomial regression, and Deep Learning (DL), which employs Long Short-Term Memory (LSTM) networks. The study makes data from the National Institute of Wind Energy (NIWE) for three years, enabling accurate energy management planning as well as long-term forecasting. In addition, advanced modelling techniques were utilized to incorporate more environmental parameters into the model to enhance prediction accuracy. AI techniques are well capable of accurate wind turbine output predictions by incorporating both linear and non-linear datasets. Moreover, this method is useful for preventive maintenance as well as estimating the potential of wind energy at new sites before the construction of wind power plants.

Keywords: Artificial Intelligence, Deep learning, Energy management, LSTM, Machine learning, Polynomial regression, Preventive maintenance, Wind turbine forecasting, Wind energy potential.

INTRODUCTION

Global environmental crises, especially around climate change, have led to a shift in humanity's energy consumption patterns from non-renewable, destructive sources to more eco-friendly renewable sources [1]. In this regard, wind energy is potentially one of the most promising renewable resources. But it is also one of the least reliable due to its dependence upon meteorological phenomena. The speed of the wind directly affects the voltage produced by the wind turbines. To integrate wind energy further into the existing forms of power generation, there is

* **Corresponding author Tiyas Sarkar:** School of Computer Science and Engineering, Lovely Professional University, Phagwara, Punjab, India; E-mail: info.tiyasofficial11901657@gmail.com

Sugandha Singh, Inam Ul Haq, Sachin Ahuja, Akib Mohi Ud Din Khanday & Neha Sharma (Eds.)

a need for additional investment in the active power and electricity market control systems to achieve the desired level of balance in the system. Such integration comes with additional expenses associated with system balancing. The most economical means of accomplishing those objectives for stakeholders like energy traders and wind farm operators is through effective unit commitment and usage, and ultimately, powerful forecasting methods [2]. Therefore, there is a need to create a forecasting system which can effectively handle the power output changes associated with medium to extreme climate change [3].

The output power of the wind turbines has been predicted using a variety of methods, primarily using a statistical approach. They are based on Betz's Law, which states that with any configuration of the wind turbine, the maximum efficiency is about 60%. Unfortunately, the availability of many nonlinear variables affecting wind power generation compounded problems for statistical methods. To address this, earlier models had to make several assumptions, which led to very big errors caused by attempting to fit linear models to such nonlinear data [4 - 10]. Such inaccuracies led to inefficiencies in responsiveness in energy management and inefficient operations of grid systems. More attention has been paid to how nonlinear systems are used in predicting wind energy, but there were no appropriate methods to manipulate data at that scale. But now, with Artificial Intelligence in the limelight, managing nonlinear data using ML and DL methods is found to be efficient. This, in conjunction with long-range forecasting models as a primary source, makes prediction of the output of wind turbines more accurate, which improves the grid planning, energy management, and the reliability of wind energy systems.

PROPOSED METHODOLOGY

The AI improvement sought was for more accurate wind turbine power output predictions, which could bring about better grid planning, energy optimization, and improved reliability of wind energy systems. The research employed different techniques of AI, such as ML and DL [11 - 20]. The results of the research were implemented by incorporating supervised learning first. A model based on polynomial regression was developed. Later on, the method employed was Long Short-Term Memory (LSTM) networks, resulting in a second predictive model. The energy management system considered in this research comprised three integral components, namely generation, distribution, and transmission. The generation segment was responsible for producing electricity, while the distribution component handled the allocation and management of energy resources. The transmission segment was tasked with delivering energy from the generation facilities to the prearranged locations.

The distribution unit was the first to communicate the projected energy efficiency or demand to the generation facilities in a 15-minute time frame. The generation units established their particular energy output based on their raw materials and workforce available during that period. Afterward, the distribution unit contacted the generation units to coordinate on the specified energy output. If the energy demand was within the output capacity of the generation plant, the transmission unit was allocated the output and managed by incorporating the output into the necessary locations. In instances of exceeding energy demand, the distribution unit solicited extra energy supply from different sources. Wind turbines were chosen in this study as one generation source in the same manner as coal-powered plants served as a backup option. Only using wind energy was not adequate; therefore, coal energy was consumed to cover the deficit. Thus, for coal energy dependence to be reduced, the accuracy of energy forecasting had to improve. This leads to better and more sustainable energy management systems.

Description of Dataset

The data set was put through training for predictive purposes by using 80% of the data for model training and 20% for model prediction tasks. This data set was created considering several alternative environmental factors, which are necessary for improving the performance prediction of turbine power. The data has been collected from the National Institute of Wind Energy (NIWE) over a span of three years, including all hourly records, which sum up to 18757 observations. From Table **1**, the dataset attributes scanned will enable us to determine the parameters affecting the turbine performance in depth for predictive analytics.

Table 1. Parameters of wind energy prediction using machine learning.

Dataset Characteristics	Detail of Description	Measurement Units
Temporal Information	Recorded in terms of hours, days, months, and years.	Not Applicable
Velocity of Airflow	Measured as the rate of air movement.	Meters per second (m/s)
Orientation of Airflow	Captures the angle of air movement.	Degrees
Horizontal Motion Component	Represents the horizontal velocity aspect	Meters per second (m/s)
Vertical Motion Component	Represents the vertical velocity aspect.	Meters per second (m/s)
Active Power Generated by the System	Power is actively utilized in kilowatts.	Kilowatts (kW)

DESCRIPTION OF WIND TURBINE

The electricity generator is intended to transform the kinetic energy of the wind into electrical energy. The main components of a wind energy converter include: the rotor, generator, and tower. The rotor is the active element in a turbine that includes blades. Its primary purpose is to transform the energy in the wind into kinetic energy in rotational form. The generator section includes an electric generator mounted with a gearbox and a number of control circuits. This device is used for the conversion of the output cable of the first set of coils and the input shaft of the second rotary device into electrical energy. The tower serves as the support structure for the rotor and generator. The rotor and the generator are mounted on the tower. The shape and the strength of the tower are also crucial factors for the turbine stability. Wind direction consists of two components: horizontal wind flow and vertical wind flow. They can be differentiated based on these components into two major groups: horizontal-axis wind turbines (HAWTs) and vertical-axis wind turbines (VAWTs). According to studies, the horizontal wind component is better most of the time.

To increase energy, the turbines are placed in such a manner that they are exposed to the wind that is available. But the power output of turbines can be affected by wind turbulence. Although turbulence takes place in short durations, it can cause issues with the condition of the blades. Hence, turbine blades are made to withstand severe wind storms and seasonal changes in winds. The gearbox takes care of torque, and the structural tower is also largely determined by the rotor. Some of the critical factors affecting power output from the turbine include the density of air, the turbine power coefficient, and the area size of the blades' swept zone. The power the turbine produces is calculated with eq. (**1**, **2**):

$$P=0.5*cp*V3*A \quad \textbf{(1)}$$

$$\text{Tip Speed Ratio} \quad \textbf{(2)}$$

All through the wind energy systems, the Power (P) produced from a wind turbine's operation depends on numerous variables. One of the components that have a major impact on this is the wind air flowing through the turbine's blades, which is expressed as velocity (V). There is also the rotor power coefficient (Cp), which defines the performance of the rotor in transforming the wind's kinetic energy into mechanical force. This is calculated by factoring in several things, such as the area (A) that the turbine blades cover when rotating into the wind. The optimization of the turbine's operational performance also rests on the critical parameter known as Tip Speed Ratio (TSR). TSR is a dimensionless number calculated from the ratio of the rotational speed of the turbine blades to the wind speed and is also significant in calculating the rotor power coefficient (Cp). It can also be expressed in the formula, which involves the radius (r) of the blade, the speed (S) of the rotor, and the wind speed (V). In a historical context, scientist Betz put forward a theoretical limit on the amount of power that can be extracted from wind using a turbine. The so-called Betz Limit states that, hypothetically, if there were an infinite number of blades, and the air conditions were ideal, then the maximum power extractable would be roughly 59.26% of the wind energy. However, in practice, this ideal condition is mostly unattainable. As a result, most wind turbines achieve about 50% power extraction efficiency. This practical limitation is called the Betz Limit, the estimated maximum power efficiency from wind energy in practice.

Process of Data Pre-processing

For a computer to interpret and handle information, it needs to be prepared in such a way that it is readable by the computer. This is necessary because computers work with machine language, which is far different from what human beings are used to. Furthermore, raw datasets are usually accompanied by some errors or discrepancies, and these errors may be of many kinds, such as human errors, device errors, or other issues. If the errors are left unchecked, they can affect the quality of the predictions made by machine learning models, resulting in incorrect outputs. Consequently, after data collection, detailed data analysis is necessary. This function will help to pinpoint any inconsistencies or problems that exist in the dataset. Once the data has been analysed, the next thing that needs to be done is data cleaning. In this phase, all the inconsistencies and mistakes that are contained in the dataset are removed so that the data is of the right quality and proper for processing. After going through the cleaning process, the next step is prepossessing. In this stage, the data is transformed into an appropriate form that can be used for predictive modelling. Depending on the nature of the data and the problem at hand, Preprocessing may include tasks such as normalization and feature selection or encoding of categorical variables.

In this case, Jupyter Software was used for both machine learning (ML) and deep learning (DL) processes. The whole undertaking was undertaken with the help of the Python programming language, which made it easier to handle data, create models, and run analyses.

Model of Machine Learning Architecture

Machine Learning (ML) is a data analytics method that is very useful when developing insights from available data. It helps machines work with old data, identify trends, and use these decisions just like people would, more or less. Among other processes, ML can be used for focusing data into some features or classes, making predictions, clustering, etc. One way of representing the processes within an ML system is by depicting them in a flowchart, which is what is done in Fig. (**1**).

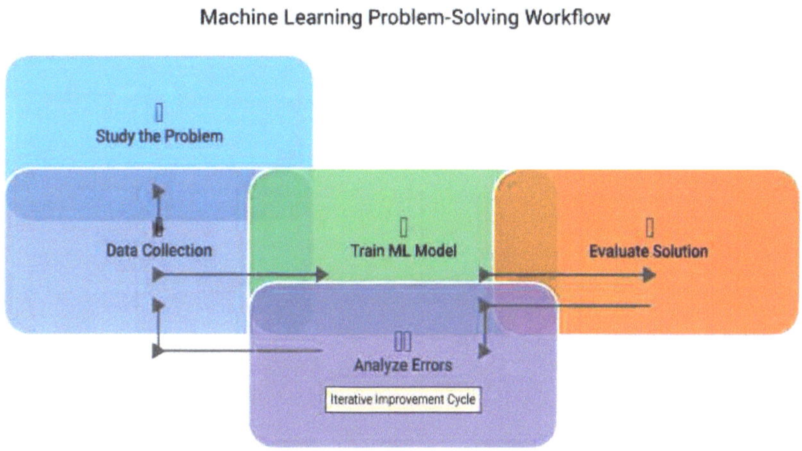

Fig. (1). To illustration flowchart of machine learning model.

ML methods are mainly divided into three broad categories, depending on the type of learning: Supervised Learning, Semi-supervised Learning, and Unsupervised Learning.

1. Supervised Learning is a learning strategy whereby the relationships of dependent and independent variables within the dataset are already comprehended. The model is fitted with the data and is expected to make predictions from the learned relationships.
2. Semi-Supervised Learning is a technique that is somewhere between supervised and unsupervised learning. Here, there is some knowledge about the dataset that is available: While certain variables are known, others are not. The model is provided with labelled data, from which it attempts to learn, while also looking for patterns in the unlabelled data.
3. Unsupervised Learning: In contrast to the other two types of ML, a model that utilizes unsupervised learning has no information about the dependent and independent variables. Here, the aim is to find patterns or structures that actually exist in the data without any prior knowledge of the variables. Fig. (**2**) demonstrates the classification of ML algorithms. The choice of a particular ML method depends on the data available and the problem that has to be solved, be it prediction, question answering, or classification. Within the scope of our dataset, the weather conditions were defined as the independent variables while the wind power output was the dependent variable. The task was to

estimate the numerical value of power generation corresponding to the given weather conditions. Since the dataset contained enough information to model the relationship between the output and the inputs, the problem was investigated using Supervised Learning and specifically Regression Analysis in order to estimate the output continuous variable from the provided input features.

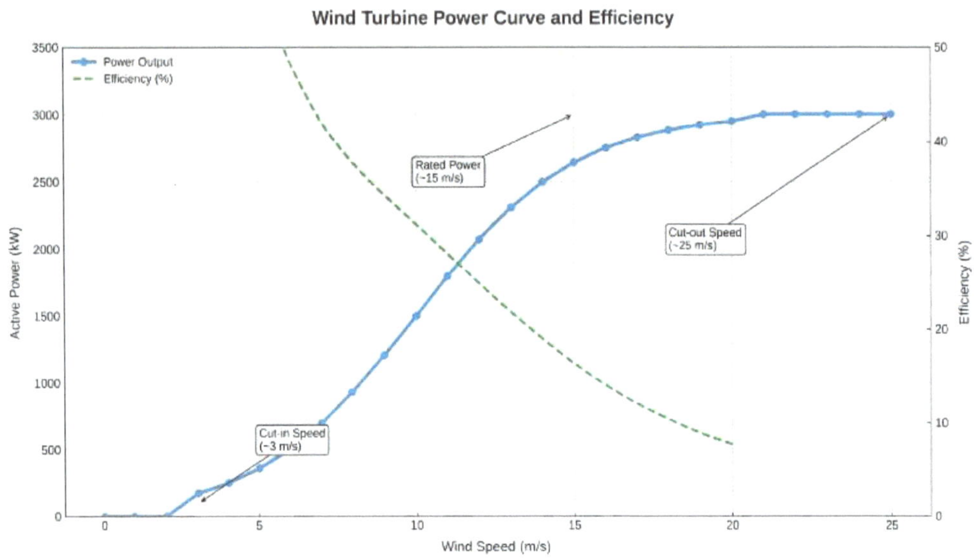

Fig. (2). Scatter plotting of wind speed and active power.

Description of Polynomial Regression

In other words, Polynomial Regression is considered a specific type of regression analysis in which the relationship between the dependent variable and the independent variable is defined as a polynomial of degree n. If we denote the dependent variable as "y", the independent variable as "x", and n as a positive integer, the equation for polynomial regression is. This type of regression is a special case of a linear regression where the model can fit a curve to the data to explain all the variations of the dependent and independent variables.

Description of Deep Learning

Deep Learning (DL) is a subdivision of Machine Learning (ML), but they are not the same. ML is a separate field, while DL models can self-modify to changing environments, making them self-regulating and self-enhancing. Like the human brain, DL builds its own neural networks, which are used for decision-making [21 - 25] and information processing, and receives feedback to improve its model.

This is the self-improving mechanism ML does not possess. As images, sounds, and symbols are complex data that both ML and DL can analyse, they both typically take a longer time to train and require a larger data set as well. In the end, DL tends to be more accurate and reliable than ML.

Description of Long-Short Term Memory

LSTM is a type of deep supervised learning that employs Recurrent Neural Networks (RNN) for data forecasts [26 - 30]. The RNN has an open-loop feedback system and works in a cycle, increasing accuracy in the model over time, making it suitable for tasks that require an understanding of time. RNNs have been shown to be effective with short-term memory, but LSTM implements long-term memory, which makes it easier to predict things such as wind turbines' performances.

PROPOSED MODEL FRAMEWORK

1st Proposed Model Analogy

The first model we created used date, time in ten-minute intervals, wind speed, wind direction, and power active from the low voltage systems as data, which was incorporated through polynomial regression. To model, we had 80% of the data available to us, while testing and prediction were done with the remaining data. The model accepts wind speed, wind direction, date, and time as the inputs, which allows it to predict power output from the wind turbine. This type of modelling can also be used to determine conditions for maintenance of the wind turbines when the result is expected to be completely different from the prediction made [31 - 33].

2nd Proposed Model Analogy

Time series data, with wind speed and wind direction as dependent variables, and power as an independent variable, was used to create the second model through the deep learning LSTM method. Once again, we used 80% of the data to train the model, while the remaining 20% was used for testing the predictions. With this model, we can maximize the efficiency of the operation of upcoming wind turbines to enable maximum wind power generation [34, 35].

RESULTS AND DISCUSSION

By utilizing polynomial regression, it came out that the mean absolute error is 19.41W, and this is shown in Fig. (**2**), which is a scatter plot showing a relation between wind speed and LV Active Power.

Fig. (**3**) shows a pie bar chart, which showcases the distribution of wind speed and direction. The prediction LSTM model shows 96% accuracy. Also, below is a comparison of predicted power output versus real power output on an hour-by-hour basis.

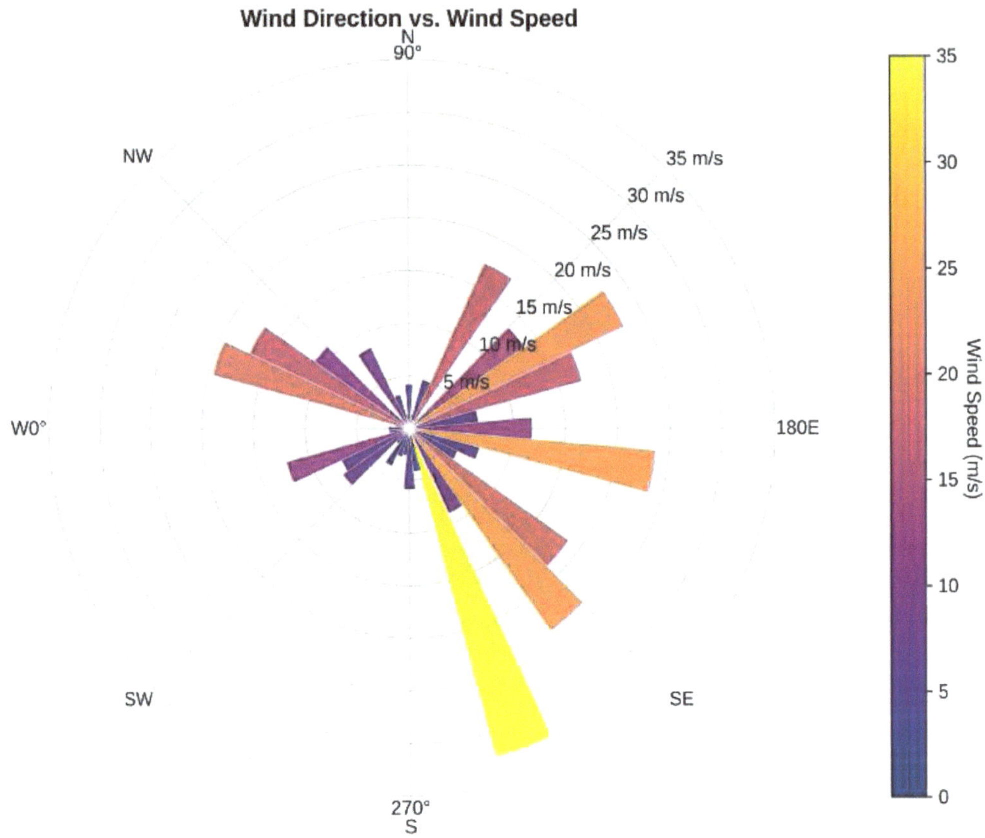

Fig. (3). Wind speed changing in a pie chart along with Wind Direction.

As shown in the Figure, the power output that the wind is predicted to have is close to the output power that will actually be supplied on an hourly basis, which is shown in Fig. (**4**).

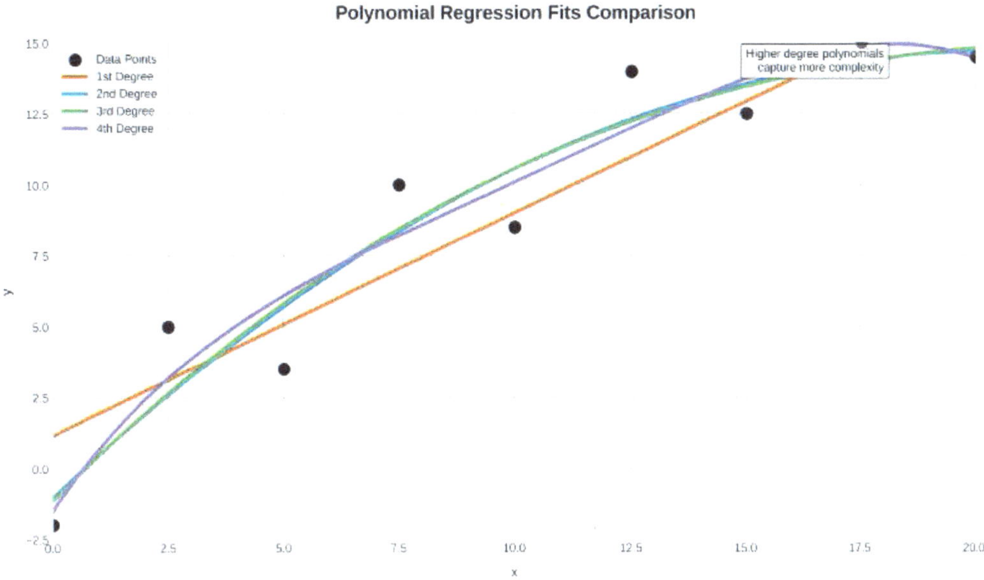

Fig. (4). Visualization output of polynomial regression.

In Fig. (**5**), there is a comparison of the power output that was observed versus the power output that was forecasted, where the blue represents the actual power values while the red indicates the predicted power.

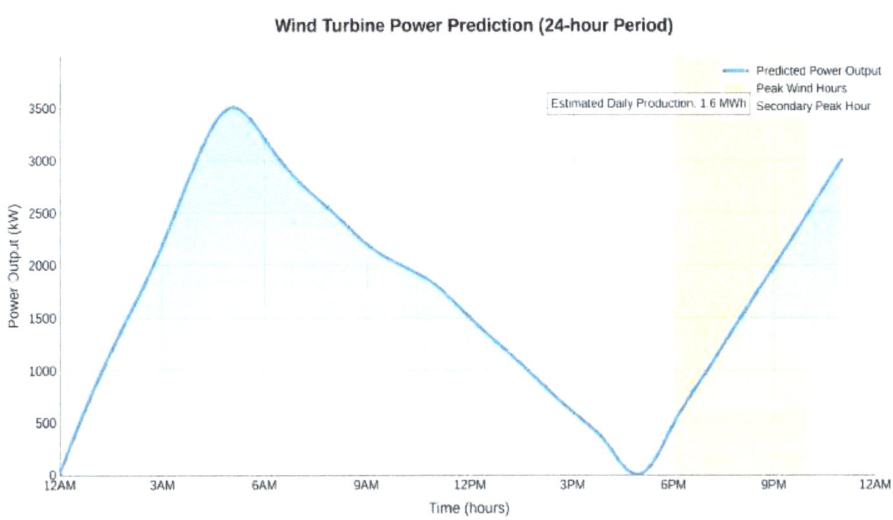

Fig. (5). The proposed model's expected *vs* actual power output.

Fig. (**6**) shows the comparison between the real and predicted treats. The data exhibits that the predictions are very close to the actual data points. The fact that this level of accuracy in power prediction is achieved is important for energy management systems. It further improves the efficiency of energy distribution. In addition, in this case, these changes could help with optimization of resource management and with lowering of environmental pollution.

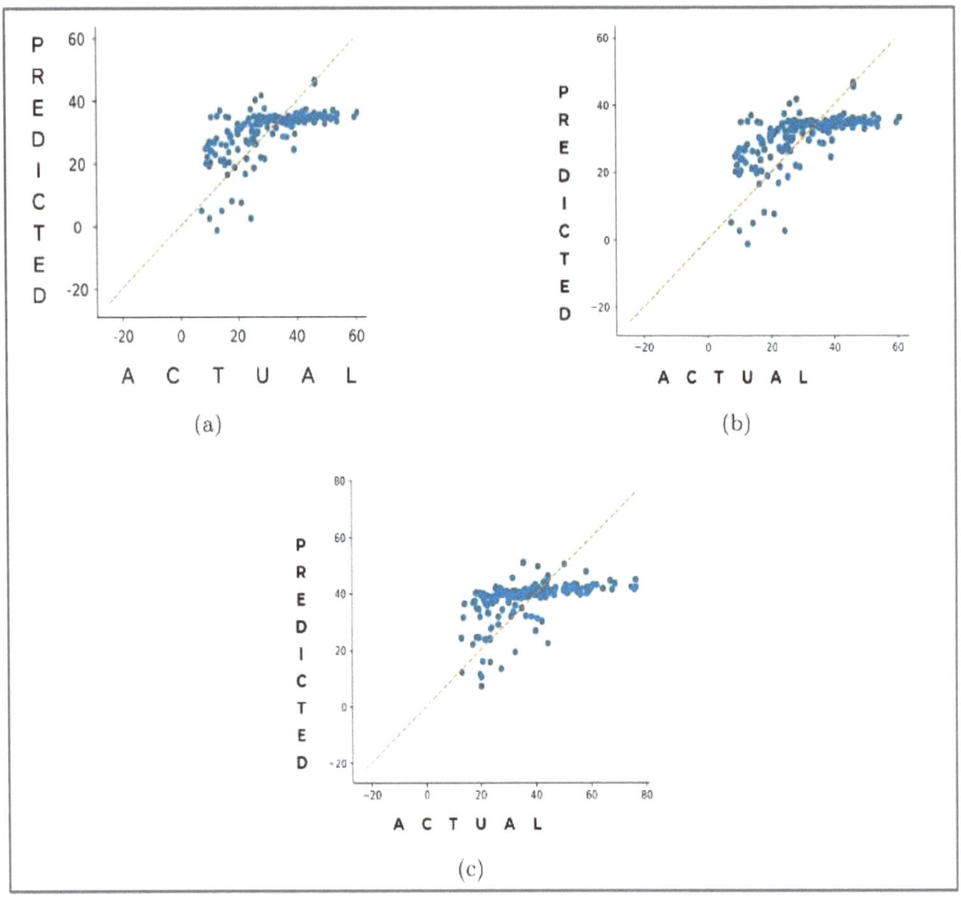

Fig. (6). Comparison graph of proposed model 2 true & predicted value.

To model, we had 80% of the data available to us, while testing and prediction were done with the remaining data. The model accepts wind speed, wind direction, date, and time as the inputs, which allows it to predict power output from the wind turbine. This type of modelling can also be used to determine conditions for maintenance of the wind turbines when the result is expected to be completely different from the prediction made. Polynomial regression is one type of regression analysis widely used in prediction that has its shortcomings. One of the drawbacks is that it is is susceptible to outliers, which affects the accuracy of the prediction values, particularly for trend lines that need to be fitted on data that is prone to outliers. While the mean absolute error that was achieved was 19.41W, this was deemed unacceptable given the sizes of the power outputs produced by wind turbines.

CONCLUSION

This undertaking seeks to solve one of the most pressing problems in the renewable energy industry: that of variability. The inherently volatile and fickle nature of renewable sources of energy, such as wind, is a major barrier to their widespread use. Because of this, a reasonable portion of their potential still goes unexploited. To address this problem, we have applied machine learning (ML) methods to improve the prediction of critical wind energy parameters, including wind speed and direction. Efficient energy forecasts are particularly important for planning and for the increasing integration of wind power into existing electricity systems. In our case, we prefer short-term to mid-term predictions, as the time frame of data collection will yield different results. As such, short-term predictions are more reliable. This also makes it easier to solve the problem at hand. The aim of the project is to employ the recently developed ML techniques in order to make the prediction processes that are already in place more effective. To begin with, we used a polynomial regression algorithm to determine the power output of wind turbines, but due to the inconsistency of winds, this approach was not very successful.

Polynomial regression is one type of regression analysis widely used in prediction that has its shortcomings. One of the drwabacks is that it is susceptible to outliers, which affects the accuracy of the prediction values, particularly for trend lines that need to be fitted on data that is prone to outliers. While the mean absolute error achieved was 19.41W, this was deemed unacceptable given the sizes of the power outputs produced by wind turbines. To mitigate these shortcomings, we used LSTM, which is an implementation of recurrent neural networks that are able to fit sequential data. In comparison to the polynomial regression, LSTM does not fail to utilize the patterns apparently present in the historical data during the predictions of the time series data. Dataset preprocessing, including cleaning and scaling, also greatly impacted the performance of the predictive model. With the intervention of LSTM, we were able to reach a very respectable prediction productivity value of 96%, indicating that LSTM has major advantages in forecasting wind energy.

ACKNOWLEDGEMENT

We extend our sincere gratitude to **Lovely Professional University (LPU)** for providing an exceptional research environment that fosters innovation and academic excellence. The university's unwavering support, access to state-of-t-e-art resources, and commitment to advancing knowledge have been instrumental in the successful completion of this research. We express our heartfelt appreciation to our mentors, colleagues, and the "Computer Science and Engineering" department at LPU for their invaluable guidance, constructive feedback, and encouragement throughout this study. Their insightful discussions and expertise have significantly contributed to refining our research. Furthermore, we acknowledge the immense support from the university's research facilities, libraries, and technical staff, whose assistance has played a vital role in shaping our findings.

REFERENCES

[1] Y. Li, P. Janik, and H. Schwarz, "Aggregated wind power characteristic curves and Artificial Intelligence for the regional wind power infeed estimation", *Electr. Eng.,* vol. 106, no. 1, pp. 655-671, 2024.
[http://dx.doi.org/10.1007/s00202-023-02005-z]

[2] C. Zhang, D. Hu, and T. Yang, "Research of artificial intelligence operations for wind turbines considering anomaly detection, root cause analysis, and incremental training", *Reliab. Eng. Syst. Saf.,* vol. 241, p. 109634, 2024.
[http://dx.doi.org/10.1016/j.ress.2023.109634]

[3] C. Biswas, A. Chakraborti, and S. Majumder, "Recent advancements in artificial intelligence and machine learning in sustainable energy management", In: *Sustainable Energy Solutions with Artificial Intelligence, Blockchain Technology, and Internet of Things.* CRC Press, 2024, pp. 35-46.

[4] X. Ding, Y. Gong, C. Wang, and Z. Zheng, "Artificial intelligence based abnormal detection system and method for wind power equipment", *Int. J. Thermofluids,* vol. 21, p. 100569, 2024.
[http://dx.doi.org/10.1016/j.ijft.2024.100569]

[5] T.A. Qureshi, and V. Warudkar, "Wind farm optimization by active yaw control strategy using machine learning approach", *Int. J. Green Energy,* vol. 21, no. 11, pp. 2628-2639, 2024.
[http://dx.doi.org/10.1080/15435075.2024.2322972]

[6] S. Chen and C. Alexiou, "Intelligent transformation and green development: a double debiased machine learning evaluation of china's IIP initiatives," *Ann. Oper. Res.*, pp. 1–33, Feb. 2026.

[7] I. Ul Haq, I. U. Hassan, and H. A. Shah, "Machine Learning Techniques for Result Prediction of One Day International (ODI) Cricket Match", In: *Proceedings of the 2023 IEEE 8th International Conference on Convergence Systems and Innovative Technologies (I2CT)*, 2023.
[http://dx.doi.org/10.1109/I2CT57861.2023.10126241]

[8] S. Bajpai, D. Sharma, and A. K. Pandey, "IoT- Enabled Healthcare Systems: Design, Implementation, and Challenges," In: *Biochip Design and Health Informatics Using IoT and SDN*, 1st ed., S. L. Tripathi, A. Gupta, A. Gupta, A. Sewak, and V. Srivastava, Eds. Hoboken, NJ: Wiley, 2026, pp. 147–166.

[9] T. Sarkar, M. Rakhra, V. Sharma, A. Maan, A. Singh, and G. S. Cheema, "Comparative study on Loan Approval Prediction Utilizing Machine Learning Approaches", In: *Proceedings of the 2024 7th International Conference on Contemporary Computing and Informatics (IC3I)*, Greater Noida, India,

2024, pp. 1521-1526.
[http://dx.doi.org/10.1109/IC3I61595.2024.10829014]

[10] H. Kaur, T. Sarkar, B. Kaur, A. Singh, G. S. Cheema and S. Das, "Using Machine Learning Techniques in Business Analytics: Lessons from Agile Project Management," In: *Proceedings of the 2024 7th International Conference on Contemporary Computing and Informatics (IC3I),* Greater Noida, India, 2024, pp. 214–220.
[http://dx.doi.org/10.1109/IC3I61595.2024.10828636]

[11] T. Sarkar, M. Rakhra, V. Sharma, A. Singh, K. Jairath and A. Maan, "Comparing Traditional vs Agile Methods for Software Development Projects: A Case Study," In: *Proceedings of the 2024 7th International Conference on Contemporary Computing and Informatics (IC3I),* Greater Noida, India, 2024, pp. 221–227.
[http://dx.doi.org/10.1109/IC3I61595.2024.10829321]

[12] S. Takkar, A. S. Rayat, V. Sharma, T. Sarkar, K. Jairath and A. Singh, "Problems and Optimal Methodologies Agile Project Management Solution in Publicly accessible Sector: A Comprehensive Evaluation of the Literature," In: *Proceedings of the 2024 7th International Conference on Contemporary Computing and Informatics (IC3I),* Greater Noida, India, 2024, pp. 207–213.
[http://dx.doi.org/10.1109/IC3I61595.2024.10828832]

[13] I. U. Haq, A. H. Rather, and G. Kaur, "A Comparative Analysis of Machine Learning Algorithms for the Early Prediction of Cardiovascular Disease," In: *Proc. 2023 2nd Int. Conf. Edge Comput. Appl. (ICECAA),* 2023, pp. 987–993.
[http://dx.doi.org/10.1109/ICECAA58104.2023.10212214]

[14] P. V. Kumari, R. K. Veluri, M. A. Nirgude, P. Pandit, A. D. Sonawane, V. Hariram, S. L. Bangare, and S. Padma, "Artificial Intelligence Techniques for Optimizing Solar and Wind Energy Production Based on Weather Patterns," *J. Mines Met. Fuels*, vol. 74, no. 1, pp. 39–47, Jan. 2026.

[15] M. A. Bhatti, C. S. Azeem, and P. Dev, "Artificial Intelligence for Predictive Maintenance and Fault Diagnosis of Wind Turbine Systems Using SCADA Data: A Comprehensive Review," *ResearchGate*, Jan. 2026.

[16] C. Lops, I. Bruno, M. Aracne, and M. Pierantozzi, "Bridging climate modeling and Artificial Intelligence for enhanced renewable energy forecasting and siting," *Renew. Energy*, vol. 262, p. 125354, Jan. 2026.
[http://dx.doi.org/10.1016/j.renene.2026.125354]

[17] G. M. Morara and E. M. Ondieki, "Leveraging artificial intelligence for renewable energy forecasting in Kenya: implications for sustainable development," Int. J. Energy Sect. Manag., Feb. 2026.
[http://dx.doi.org/10.1108/IJESM-08-2025-0036]

[18] S. Kumar, V. K. Sharma, D. K. Patel, T. Sagar, and G. Nandan, "Artificial Intelligence-Driven Flow Optimization for Enhanced Efficiency in Renewable Energy Systems," In: *Artificial Intelligence and Computational Modeling in Heat Transfer and Fluid Dynamics*, 1st ed., M. K. Awasthi and R. G. Agarwal, Eds. Hoboken, NJ: Wiley-Scrivener, 2026, pp. 277–306.
[http://dx.doi.org/10.1002/9781394433605.ch10]

[19] S. A. Soomro, N. H. Mirjat, M. A. Uqaili, N. Arshad, and S. A. Khatri, "Wind Speed Prediction Using Sparse SCADA Data of an Offshore Wind Farm of Pakistan Using Different Machine Learning Techniques," *Results Eng.*, vol. 26, p. 109144, June 2026.
[http://dx.doi.org/10.1016/j.rineng.2026.109144]

[20] D. Ewim, A. O. Eweje, O. A. Hamza, A. S. Collins, and E. C. Onukwulu, Data Analytics and Business Intelligence: AI-Driven Solutions and Energy Applications, 1st ed. London, UK: Cambridge Scholars Publishing, 2026.

[21] D. Ewim, A. O. Eweje, O. A. Hamza, A. S. Collins, and E. C. Onukwulu, *Data Analytics and Business Intelligence: AI-Driven Solutions and Energy Applications*, 1st ed. London, UK: Cambridge Scholars Publishing, 2026.

[http://dx.doi.org/10.1109/IC3SE62002.2024.10593475]

[22] A. H. Rather, and I. U. Haq, "Intelligent Framework for Early Prediction of Diabetic Retinopathy: A Deep Learning Approach", *Proceedings of International Conference on Paradigms of Communication, Computing and Data Analytics PCCDA 2023*, A. Yadav, S. J. Nanda, and M. H. Lim, Eds., Singapore: Springer, 2023, pp. 315–325. [http://dx.doi.org/10.1007/978-981-99-4626-6_32]

[23] P. Chakrabarty, T. Sarkar, M. Rakhra, K. Jairath and V. Sharma, "Enhanced Data Security Framework Using Lightweight Cryptography and Multi-Level Encryption," 2024 International Conference on Communication, Computer Sciences and Engineering (IC3SE), Gautam Buddha Nagar, India, 2024, pp. 720-725. [http://dx.doi.org/10.1109/IC3SE62002.2024.10593191]

[24] T. Sarkar, M. Rakhra, V. Sharma and A. Singh, "An Empirical Comparison of Machine Learning Techniques for Bank Loan Approval Prediction," 2024 International Conference on Communication, Computer Sciences and Engineering (IC3SE), Gautam Buddha Nagar, India, 2024, pp. 137-143. [http://dx.doi.org/10.1109/IC3SE62002.2024.10593355]

[25] S. Talwani, M. Rakhra and T. Sarkar, "Perspectives on the Future of Agriculture and Recent Developments in Agricultural Technology," 2024 11th International Conference on Reliability, Infocom Technologies and Optimization (Trends and Future Directions) (ICRITO), Noida, India, 2024, pp. 1-6. [http://dx.doi.org/10.1109/ICRITO61523.2024.10522413]

[26] T. Sarkar, B. Moharana, M. Rakhra and G. S. Cheema, "Comparative Analysis of Empirical Research on Agile Software Development Approaches," 2024 11th International Conference on Reliability, Infocom Technologies and Optimization (Trends and Future Directions) (ICRITO), Noida, India, 2024, pp. 1-6. [http://dx.doi.org/10.1109/ICRITO61523.2024.10522134]

[27] A. K. Verma, T. Sarkar, M. Rakhra, V. K. Pandey and P. Chakrabarty, "Studying the Impact of Voice-Activated Chatbot's with Artificial Intelligence on Consumer Interaction in India," 2024 4th International Conference on Technological Advancements in Computational Sciences (ICTACS), Tashkent, Uzbekistan, 2024, pp. 1708-1713. [http://dx.doi.org/10.1109/ICTACS62700.2024.10841074]

[28] Moharana, V. K. Singh, T. Sarkar, A. Hassan and V. K. Pandey, "Predicting Indian Master's and PhD Enrollment using Data Mining & Machine Learning Approaches," 2024 4th International Conference on Technological Advancements in Computational Sciences (ICTACS), Tashkent, Uzbekistan, 2024, pp. 1714-1720. [http://dx.doi.org/10.1109/ICTACS62700.2024.10841260]

[29] T. Sarkar, M. Rakhra, P. Singh, P. K. Angra, G. S. Cheema and N. Mathur, "Comprehensive Review on Fog Computing: Obstacles, Feature Traits, Limitations, and Potential Leverages on Application," 2024 4th International Conference on Technological Advancements in Computational Sciences (ICTACS), Tashkent, Uzbekistan, 2024, pp. 404-410. [http://dx.doi.org/10.1109/ICTACS62700.2024.10840690]

[30] Rathod, Umang H., Vinayak Kulkarni, and Ujjwal K. Saha. "On the application of machine learning in savonius wind turbine technology: an estimation of turbine performance using artificial neural network and genetic expression programming." *J. Energy Resour. Technol.*, vol. 144, no. 6, Art. no. 061301, Jun. 2022. [http://dx.doi.org/10.1115/1.4051736]

[31] I. Ul Haq, "Diagnosis of Neurological Disease Using Bioinspired Algorithms", In: Bio-Inspired Optimization for Medical Data Mining, S. Srivastava, A. Anand, A. Kumar, B. Saini, P. S. Rathore, Eds., *Energies*, vol. 14, no. 7, Art. no. 1845, Mar. 2021. pp. 245-270. [http://dx.doi.org/10.1002/9781394214211.ch12]

[32] A.N. Sharkawy, A. Ameen, S. Mohamed, G. Abdel-Jaber, and I. Hamdan, "Design, assessment, and

modeling of multi-input single-output neural network types for the output power estimation in wind turbine farms", *Automation,* vol. 5, no. 2, pp. 190-212, 2024. [http://dx.doi.org/10.3390/automation5020012]

[33] H. R. F. Fotso, C. V. A. Kazé, and G. D. Kenmoé, "Real-time rolling bearing power loss in wind turbine gearbox modeling and prediction based on calculations and artificial neural network", *Tribol. Int.,* vol. 163, p. 107171, 2021. [http://dx.doi.org/10.1016/j.triboint.2021.107171]

[34] A. Santolamazza, D. Dadi, and V. Introna, "A Data-Mining Approach for Wind Turbine Fault Detection Based on SCADA Data Analysis Using Artificial Neural Networks", *Energies,* vol. 14, no. 7, p. 1845, 2021. [http://dx.doi.org/10.3390/en14071845] [http://dx.doi.org/10.3390/en14071845]

[35] T. Sarkar, A. Bhattacherjee and M. Mukhia, "Large Language Model for Practical Evaluation in Cloud Computing with Applications of Real World Data," In: *Proceedings of the 2024 4th International Conference on Technological Advancements in Computational Sciences (ICTACS),* Tashkent, Uzbekistan, 2024, pp. 1721–1729. [http://dx.doi.org/10.1109/ICTACS62700.2024.10840404]

CHAPTER 10

The Future of Resource Optimization with AI and Machine Learning

Khalid Hafiz Mir[1,*], **Anzah Bashir**[1] and **Hashim Mohi u din Dar**[2]

[1] *School of Computer Science and Technology, Lovely Professional University, Phagwara, Punjab, India*

[2] *Department of Computer Science & Engineering, Rayat Bahra University, Mohali, Punjab, India*

Abstract: Artificial Intelligence (AI) and Machine Learning (ML) have been making rapid progress that has revolutionized the optimization of resources across various industries. These technologies allow for smarter, more efficient decision-making. This paper explores the transformative role of AI and ML in optimizing resources such as energy, computing power, logistics, and human capital. AI-driven systems enhance efficiency, reduce waste, and improve cost-effectiveness through predictive analytics, real-time monitoring, and adaptive algorithms. The key applications are AI-powered energy grids, intelligent supply chain management, dynamic cloud resource allocation, and workforce optimization. The combination of AI with edge computing and IoT further increases responsiveness and scalability. However, challenges such as data privacy, ethical considerations, and computational complexity must be addressed to maximize AI's potential. This study shows emerging trends, real-world applications, and future research directions that lead to AI-driven resource optimization toward sustainable and intelligent resource management.

Keywords: Autonomous systems, Energy management, Internet of things, Predictive analytics, Quantum computing, Resource optimization, Real-time decision making, Sustainability, Supply chain management, Smart cities.

INTRODUCTION

In today's ever-changing technological environment and increasingly constrained resources, the proper utilization and allocation of available resources is one of the key problems in modern industrial and organizational life. The traditional approaches to managing resources use static models, human experience, and history; all these approaches are no longer accurate for the dynamics of complex systems that are now dealing with modern dynamic systems. The integration of

* **Corresponding author Khalid Hafiz Mir:** School of Computer Science and Technology, Lovely Professional University, Phagwara, Punjab, India; E-mail: mamta.airesearch@gmail.com

Sugandha Singh, Inam Ul Haq, Sachin Ahuja, Akib Mohi Ud Din Khanday & Neha Sharma (Eds.)

Artificial Intelligence (AI) and Machine Learning (ML) has emerged as a transformative solution, enabling data-driven, adaptive, and highly efficient resource optimization across multiple domains [1].

Through AI and ML technology, the process of decision making allows evaluation of large datasets to discover valuable insights, and in turn, an optimal strategy and instant response [2]. Because these technologies maintain sustainable operations while achieving greater efficiency and reduced costs, their application has proven successful in several specialized industries. Modern resource management depends greatly on AI-driven models, which use dynamic process optimization to predict demand, detect inefficiencies, and execute continuous performance improvements.

THE EVOLUTION OF RESOURCE OPTIMIZATION

Organizations and industries have spotlighted resource optimization as a serious concern that they have been addressing for many centuries. The technological tools for resource optimization have experienced major changes throughout history since technological advancements, scientific developments, and worldwide economic pressures took hold [3]. The journey towards resource optimization is a story of continuous evolution, beginning with the use of manual methods and static models by humans and continuing with AI-powered and data-driven automated solutions today.

Traditional Methods

The essential nature of resource optimization throughout history led people to develop different early resource allocation methods that optimized their available resources. Traditional resource optimization techniques showed weaknesses due to both slow operation and inability to respond quickly to sudden developments [4]. Traditional methods have three fundamental approaches:

- **Linear Programming:** Linear programming is a mathematical technique that helps users get the best outcomes from certain mathematical frameworks.
- **Heuristics:** By employing heuristic solutions, organizations can get general techniques for attaining desirable results.
- **Simulation Models:** A computational technique for simulating real-world scenarios is the simulation model.

Scientists implemented these procedures effectively at that time, but these methods became less effective because of complex mathematical calculations that needed additional empirical data promptly.

The Digital Transformation

Digital technology has established a substantial change for optimizing resource utilization. Through computers, the possibility of advanced calculations and data evaluation improved both accuracy and efficiency of resource allocation. These tools included Enterprise Resource Planning (ERP) systems alongside advanced analytics solutions, which started becoming mainstream practices [5].

The AI Revolution

AI, together with ML, operates as the future advancement in resource optimization technology. Modern technologies allow users to handle enormous datasets alongside pattern identification and accurate prediction making at exceptional speed [6]. AI is revolutionizing resource management through autonomous decision-making and predictive analytics.

Understanding AI and Machine Learning

The fundamental problem-solving approach in every industry moves forward radically with Artificial Intelligence (AI) and Machine Learning (ML) systems, particularly in resource optimization methods. Systems utilize these technologies to gain knowledge from data so they can independently choose optimized decisions to enhance their operational performance, as shown in Fig. (**1**).

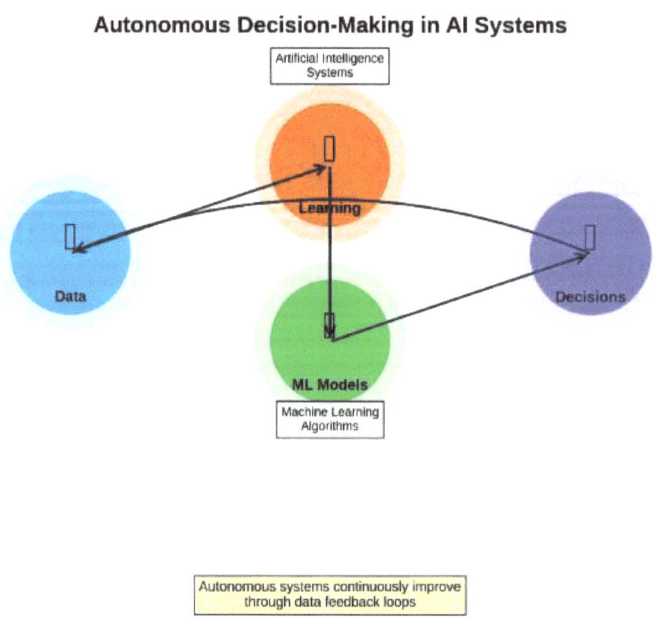

Fig. (1). AI and ML hierarchy.

WHAT IS ARTIFICIAL INTELLIGENCE (AI)?

Computers designed through AI possess the capability to handle jobs that need human intelligence to complete [7]. Computers perform three main intelligent activities: identifying patterns, making choices, and solving problems. Two major types make up the classification of AI.

Narrow AI: Specific AI systems like Siri and Alexa, along with recommendation engines like Netflix and Amazon, fall under the category of Narrow AI.

General AI: General AI describes theoretical machines with human intellectual capabilities to solve any intellectual challenge.

Through its automation capabilities, AI both reduces human labor requirements and enhances the ability to make decisions across industries that include health care operations, as well as financial sectors, together with logistical operations.

WHAT IS MACHINE LEARNING (ML)?

Machine Learning operates as part of AI by teaching computers to understand different patterns in provided datasets through unprogrammed learning methods. ML algorithms do not operate by following predetermined rules, but they process data to detect patterns, which allows performance enhancement through time [8].

Machine learning exists in three specific forms.

Supervised Learning: The algorithm achieves its learning process by analyzing data that is properly identified. The detection system for spam emails applies supervised learning through "spam" and "not spam" email labels.

Unsupervised Learning: The algorithm analyzes unlabeled data to establish hidden patterns in an unsupervised approach. Example: Customer segmentation in marketing.

Reinforcement Learning: The algorithm achieves learning through direct interaction with its environment when it receives positive outcomes for making beneficial choices. The implementation of self-driving cars results in better abilities of human drivers.

APPLICATIONS OF AI AND ML IN RESOURCE OPTIMIZATION

Artificial Intelligence, together with Machine Learning, enables different industries to achieve total resource optimization improvements through its implementation. Organizations obtain real-time predictions through large data processing methods that lead to cost reductions as well as resource optimization and operational efficiency improvements. Here are some key applications, as shown in Fig. (**2**):

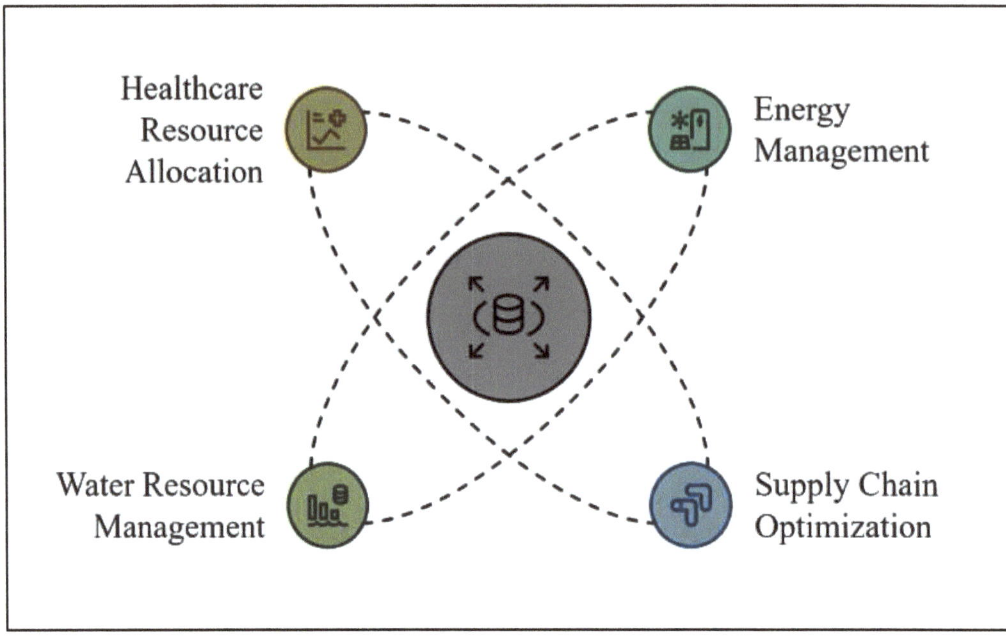

Fig. (2). AI and ML in resource optimization.

Energy Management

Smart energy grids, along with factory systems, make use of AI predictive power to enhance power consumption by monitoring patterns and then modify supply behavior to achieve exceptional efficiency levels.

Supply Chain Optimization

The systematic process of supply chain optimization creates better corporate resource distribution with enhanced communication for reduced expenses and advanced performance results [9]. This is a system that delivers products efficiently while decreasing waste and minimizing delivery times. The effective market prediction combined with stock control and movement optimization requires both Artificial Intelligence and data analytics systems from businesses. Businesses that optimize their supply chains receive three primary advantages: decreasing costs across multiple stages while delivering products faster with more satisfied customers.

Water Resource Management

AI techniques improve water management at agricultural fields alongside industrial establishments and urban settlement areas. The combination of data consisting of weather patterns, soil conditions, and water consumption patterns allows machine learning models to achieve accurate demand predictions. Artificial Intelligence-based systems that monitor water leaks will help manage water distribution effectively as they track down water losses.

Healthcare Resource Allocation

Hospital bed administration and staff, as well as equipment management, improve through Artificial Intelligence systems working with machine learning to optimize medical resources. Patient data evaluation through predictive models determines needs forecasting that helps hospitals prepare for emergencies [10]. The combination of smart inventory control and AI scheduling platforms delivers critical equipment promptly to minimize health service losses while improving medical care quality.

CASE STUDIES: SUCCESS STORIES

The utilization of real-life scenarios demonstrates to industries how Artificial Intelligence, teamed with machine learning, delivers efficient resource optimization. AI successes in organizations yield better performance and reduced costs alongside outstanding decision-making capabilities [11]. The examination of real-life case studies helps operators of businesses and research organizations to recognize specific beneficial outcomes AI offers when solving complex problems. Research papers document complete information regarding AI/ML techniques that resolve specific problems through their systematic achievement methods. Six vital case studies appear in Table **1**.

Table 1. Key AI-driven optimizations across different industries.

Case Study	Company	Industry	AI/ML Application	Impact	Date
Google's Data Center Cooling	Google	Technology	AI optimizes cooling systems in data centers.	40% reduction in energy consumption.	2016
Walmart's Inventory Management	Walmart	Retail	ML predicts product demand and optimizes inventory.	Reduced waste and improved customer satisfaction.	2017
IBM's AI for Water Management	IBM	Agriculture	AI analyzes satellite and weather data for irrigation.	Significant water savings and increased crop yields.	2019

(Table 1) cont.....

Case Study	Company	Industry	AI/ML Application	Impact	Date
Amazon's Supply Chain Optimization	Amazon	E-commerce	AI automates warehouse operations and logistics.	Faster deliveries and lower operational costs.	2020
Tesla's Autonomous Fleet Management	Tesla	Automotive	AI manages self-driving technology and fleet efficiency.	Improved traffic management and fuel efficiency.	2021
Siemens' Smart Energy Management	Siemens	Energy	AI optimizes power grid operations and energy distribution.	Increased efficiency and lower energy costs.	2022

CHALLENGES AND ETHICAL CONSIDERATIONS

AI and machine learning progress demands attention towards multiple obstacles and moral issues while maintaining effective, responsible implementation.

Data Privacy

The efficient performance of AI and ML depends strictly on having big data sets. The data privacy and security concerns of sensitive information become central issues because of this situation [12]. Users require organizations to establish robust data protection standards alongside GDPR compliance to prevent unauthorized data access and improper use.

Bias in Algorithms

AI systems extract their knowledge from previous datasets that potentially hold discriminatory elements [13]. Unmonitored AI systems tend to duplicate existing gender or racial or socio-economic biases as they make decisions that may prove unfairly discriminatory, as described in Table **2**. The reduction of bias requires developers to build training datasets that represent diverse populations, perform regular audits, and implement systematic methods against biases.

Table 2. Summary of how AI, IoT, sustainability efforts, and quantum computing will shape the future of resource optimization.

Type of Bias	Usage in AI (%)	Description	Example	Impact	Current Applications in AI
Selection Bias	35%	Data used to train AI is not representative of all groups.	Facial recognition systems are failing to detect darker skin tones.	Discrimination in security and hiring systems.	Facial recognition, hiring tools, predictive policing

(Table 2) cont.....

Type of Bias	Usage in AI (%)	Description	Example	Impact	Current Applications in AI
Confirmation Bias	25%	AI reinforces existing beliefs, lacking diverse perspectives.	Social media feeds only show content similar to past interactions.	Increases political and social polarization.	News recommendation engines, search algorithms
Automation Bias	20%	Users overly trust AI decisions without human verification.	AI in healthcare misdiagnoses a condition without doctor review.	It can lead to life-threatening mistakes in medicine.	AI-driven medical diagnosis, autonomous vehicles
Labeling Bias	10%	Errors in labeling training data due to human subjectivity.	Sentiment analysis misclassifies sarcasm as positive sentiment.	AI misinterprets intent, affecting customer support.	Sentiment analysis, spam detection
Historical Bias	10%	AI learns from past biased data, reinforcing inequality.	Loan approval models favor wealthier neighborhoods.	Perpetuates social inequalities in finance and justice.	Loan approvals, criminal justice risk assessments

Job Displacement

When organizations automate all repetitive operations, they might put workers at risk of job elimination in manufacturing and logistics fields and customer service roles. New positions that have emerged due to AI advancements exist in areas of Artificial Intelligence development, together with system monitoring and cybersecurity [14-16]. Leadership at both the workplace and government level need to establish training initiatives for retraining workers into AI-relevant positions.

Ethical AI

AI technology needs development practices that enable transparency, along with fairness and accountability during its utilization. The execution of AI makes three fundamental ethical challenges visible due to decision-making responsibilities and AI misuse threats (deepfakes and surveillance), as well as the requirement that AI benefits multidimensional human interests without human exploitation [17]. The framework for ethical AI practices needs to implement proper AI deployment procedures alongside human supervision during algorithmic decision-making processes.

Future Trends and Predictions

This evolution of AI alongside machine learning technologies will establish different developing trends that will define resource optimization in the future across all industries, as shown in Fig. (**3**). The emerging developments will create more efficient operations, sustainable practices, and improved decision support systems.

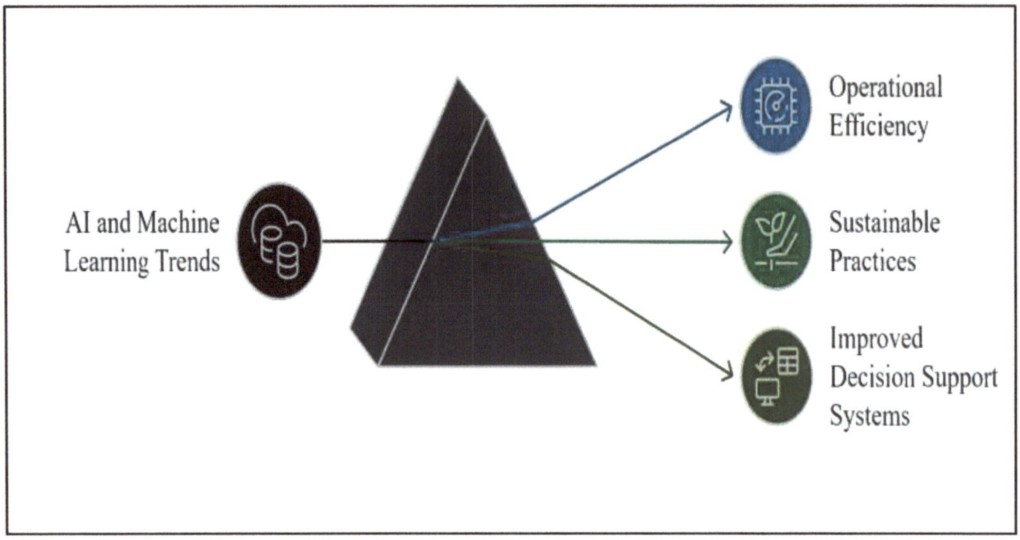

Fig. (3). Unveiling AI's future in resource optimization.

Autonomous Systems

The implementation of autonomous systems will enable AI technology to exploit resources through unattended operations. The systems will use their ability to learn as they operate to make ongoing improvements, which result in real-time operational enhancements for manufacturing and energy consumption.

Integration with IoT

AI operates as a partner to the Internet of Things (IoT) system to optimize resources in real time through connected devices. The analysis of multiple connected devices enables smart cities to maximize their traffic flows, energy use and waste management operations through AI systems.

AI-DRIVEN SUSTAINABILITY

AI technology stands vital in sustainability because it optimizes resource management to reduce waste output. The technology supports renewable energy management systems along with carbon footprint reduction and supports industrial circular economy processes.

QUANTUM COMPUTING

Through quantum computing, AI achieves enhanced operational abilities to solve complex optimization tasks with enhanced speed capabilities. The speedup in data processing and optimization across logistics, energy management, and various industries is possible through these developments.

The evolution of AI shows promise with trends that help increase efficiency while enhancing sustainability measures and decision-making capabilities in resource optimization procedures.

CONCLUSION

The union of AI with Machine Learning in resource optimization establishes an entirely new method for managing resource use in our current era. Through their implementation, these technologies enable efficiency breakthroughs that combine sustainability with strong profitability models. Realizing these technological potential demands attention to critical obstacles, which involve data privacy issues, algorithmic bias, and ethical concerns. We ought to push forward through the exploration of AI and ML capabilities alongside responsible and ethical usage management of these technologies. Resource optimization remains in an upward trend because AI and ML technology will dominate the path toward its future development.

REFERENCES

[1] M. Soori, B. Arezoo, and R. Dastres, "Artificial intelligence, machine learning and deep learning in advanced robotics, a review", *Cognitive Robotics,* vol. 3, pp. 54-70, 2023. [http://dx.doi.org/10.1016/j.cogr.2023.04.001]

[2] I. Ul Haq, I. U. Hassan, and H. A. Shah, "Machine Learning Techniques for Result Prediction of One Day International (ODI) Cricket Match", In: *Proceedings of the 2023 IEEE 8th International Conference on Convergence Systems and Innovative Technologies (I2CT)*, 2023. [http://dx.doi.org/10.1109/I2CT57861.2023.10126241]

[3] S.B. Junaid, A.A. Imam, A.O. Balogun, L.C. De Silva, Y.A. Surakat, G. Kumar, M. Abdulkarim, A.N. Shuaibu, A. Garba, Y. Sahalu, A. Mohammed, T.Y. Mohammed, B.A. Abdulkadir, A.A. Abba, N.A.I. Kakumi, and S. Mahamad, "Recent advancements in emerging technologies for healthcare management systems: a survey", *Healthcare (Basel),* vol. 10, no. 10, p. 1940, 2022. [http://dx.doi.org/10.3390/healthcare10101940] [PMID: 36292387]

[4] C. Yaiprasert, and A.N. Hidayanto, "AI-powered ensemble machine learning to optimize cost strategies in logistics business", *Int. J. Inf. Manag. Data Insights,* vol. 4, no. 1, p. 100209, 2024. [http://dx.doi.org/10.1016/j.jjimei.2023.100209]

[5] A.Y. Zakaria, E. Abdelbadea, A. Raslan, T. Ali, M. Gheith, A.S. Khater, and E.A. Amin, "An Improved Enterprise Resource Planning System Using Machine Learning Techniques", *J. Softw. Eng. Appl,* vol. 17, no. 5, pp. 203-213, 2024.

[http://dx.doi.org/10.4236/jsea.2024.175011]

[6] Q. Bndyan, R. Salih, K. A. A. Ali, K. M. Mousa, and D. H. Ahmed, "A neural network-based framework for enterprise financial error correction using AI and big data," *Sci. Rep.*, vol. 16, no. 1, p. 48510, Apr. 2026.
[http://dx.doi.org/10.1016/j.inffus.2023.101945]

[7] I. U. Haq, A. H. Rather, and G. Kaur, "A Comparative Analysis of Machine Learning Algorithms for the Early Prediction of Cardiovascular Disease", In: *Proceedings of the 2023 2nd International Conference on Edge Computing and Applications (ICECAA)*, 2023, pp. 987-993.
[http://dx.doi.org/10.1109/ICECAA58104.2023.10212214]

[8] A.H. Rather, and I.U. Haq, "Intelligent Framework for Early Prediction of Diabetic Retinopathy: A Deep Learning Approach", In: *Proceedings of International Conference on Paradigms of Communication, Computing and Data Analytics: PCCDA 2023,* 2023. Singapore: Springer, 2023, pp. 377–388.
[http://dx.doi.org/10.1007/978-981-99-4626-6_32]

[9] A. Elbeltagi, A. Srivastava, N.L. Kushwaha, C. Juhász, J. Tamás, and A. Nagy, "Meteorological data fusion approach for modeling crop water productivity based on ensemble machine learning", *Water,* vol. 15, no. 1, p. 30, 2022.
[http://dx.doi.org/10.3390/w15010030]

[10] I. Ul Haq, "Diagnosis of Neurological Disease Using Bioinspired Algorithms", In: *Bio-Inspired Optimization for Medical Data Mining*, S. Srivastava, A. Anand, A. Kumar, B. Saini, P. S. Rathore, Eds. Hoboken, NJ, USA: Wiley-IEEE Press, 2024, pp. 245–270.
[http://dx.doi.org/10.1002/9781394214211.ch12]

[11] M. Soori, F. K. G. Jough, R. Dastres, and B. Arezoo, "AI-based decision support systems in Industry 4.0, a review", *J. Econ. Technol.,* vol. 4, pp. 206-225, 2026.
[http://dx.doi.org/10.1016/j.ject.2024.08.005]

[12] D. Sargiotis, "Data Security and Privacy: Protecting Sensitive Information", In: *Data Governance.* Springer Nature Switzerland: Cham, 2024, pp. 217-245.
[http://dx.doi.org/10.1007/978-3-031-67268-2_6]

[13] M. Veale, and R. Binns, "Fairer machine learning in the real world: Mitigating discrimination without collecting sensitive data", *Big Data Soc.,* vol. 4, no. 2, 2017.
[http://dx.doi.org/10.1177/2053951717743530]

[14] I. Batra, "Luhach AKr, Pathak N. Research and Analysis of Lightweight Cryptographic Solutions for Internet of Things", *Proceedings of the Second International Conference on Information and Communication Technology for Competitive Strategies,* ACM: New York, NY, USA, pp. 1-5, 2016.

[15] Z. D. Turay, M. Koroma, M. S. Fofanah, and D. B. Kaitibi, "AI-Driven Cybersecurity in Industry 4.0: A Global Comparative Analysis of Detection Methods, Challenges, and Resilient Solutions," *Int. J. Adv. Res.*, vol. 14, no. 1, pp. 54–75, Jan. 2026.
[http://dx.doi.org/10.1016/j.dim.2023.100063]

[16] S. S. Mohammed, H. I. Mohammed, N. S. Mohammed, N. M. Shafiq, and H. Q. Flayyih, "Future Directions in Artificial Intelligence for Cybersecurity: Emerging Trends and Key Developments," *Passer J. Basic Appl. Sci.*, vol. 8, no. 1, pp. 400–409, Feb. 2026.
[http://dx.doi.org/10.1016/j.inffus.2023.101804]

[17] C. Sekhar Veluru, "Responsible Artificial Intelligence on Large Scale Data to Prevent Misuse, Unethical Challenges and Security Breaches", *J. Artif. Intell. Cloud Comput.,* vol. 3, no. 1, pp. 1-6, 2024.
[http://dx.doi.org/10.47363/JAICC/2024(3)167]

CHAPTER 11

Optimizing Wind Farm Site Selection with AI

Vandita Nandal[1], **Himani Uppal**[1], **Syed Zoofa Rufai**[1,*] and **Inam Ul Haq**[2]

[1] *Department of Computer Science and Engineering (CSE), Shree Guru Gobind Singh Tricentenary University (SGT) University, Gurugram, Haryana, India*

[2] *School of Engineering and Technology (SET), CGC University, Mohali, Punjab, India*

Abstract: Wind energy is becoming more and more important as a sustainable and eco-friendly power source due to the growing need for renewable energy. Wind farm site selection plays a very important role in energy generation, also minimizing the factors that restrict both economic activity and environmental protection. Traditional methods used for wind farm site selection depend on manual data collection, studies, and research, which are more time-consuming, have data limitations, and are difficult to manage. Artificial Intelligence (AI) has become a revolutionary solution, and it also enhances the efficiency and accuracy of the wind farm site selection process. AI algorithms such as Artificial Neural Networks (ANN), Support Vector Machines (SVMs), Random Forest, Genetic Algorithms (GAs), and Deep Learning (DL) enable more accurate wind resource assessment, turbine placement, and other factors affecting the wind farm site selection process. In future hybrid AI models, cloud computing, and Explainable AI (XAI) aim to ensure more transparent and reliable wind farm site selection. By integrating AI technologies, the wind energy sector can gain higher efficiency and greater sustainability, which supports the global transition to renewable energy.

Keywords: Artificial Intelligence, Artificial Neural Network (ANN), Cloud computing, Deep learning, Economic activity, Explainable AI (XAI), Global transition, Revolutionary, Renewable energy, Support Vector Machine (SVM), Sustainable.

INTRODUCTION

The Growing Importance of Wind Energy

Renewable energy harnesses the energy of the wind to produce electricity. Wind turbines convert the energy of moving air into electrical energy. As the wind blows over the blades of a turbine, the force of the wind causes the turbine blades

* **Corresponding author Syed Zoofa Rufai:** Department of Computer Science and Engineering (CSE), Shree Guru Gobind Singh Tricentenary University (SGT) University, Gurugram, Haryana, India; E-mail: syedzoofa78@gmail.com

Sugandha Singh, Inam Ul Haq, Sachin Ahuja, Akib Mohi Ud Din Khanday & Neha Sharma (Eds.)

to revolve, powering a generator that creates energy. Since this energy-generating technology produces no greenhouse gases or air pollutants, it has been considered an environmentally friendly technology for the future. Harnessing renewable energy, such as wind and solar energy, is very important to slow down the greenhouse gas consequences and to reduce our reliance on fossil fuels [1].

Wind energy is a clean, renewable resource that meets the growing demand for sustainable power. As environmental deterioration becomes a bigger problem, governments, businesses, and consumers are increasingly using renewable energy sources to meet their energy needs. The world's transition to a more sustainable energy future depends heavily on wind energy because of its enormous potential to produce electricity with little negative impact on the environment.

Wind energy is a good alternative to fossil fuels and has been used for a long time. And it is beneficial in that it is very abundantly available, unlimited, and sustainable. The number of wind farms being constructed worldwide has steadily increased during the last ten years. According to the World Wind Energy Association (WWEA), over 52 GW of capacity was built in 2017, bringing the total capacity to 539 GW globally [2]. The integration cost of wind power into the power grid is low [3]. Wind energy is also an important tool in the diversification of the energy mix, which serves as an alternative to the old-school fossil fuel-based power generation. Countries can rely less on coal, oil, and natural gas substitutes through wind power generation. Thus, wind energy contributes to alleviating the adverse effects of greenhouse gas emissions but also serves the purpose of enhancing energy security through the offer of a more reliable and diverse energy supply. The ability of wind energy systems to generate electricity at distant, remote, or offshore locations offers a unique opportunity for supplying electricity to regions that rarely connect with conventional energy infrastructures.

In addition, technological advancements in manufacturing have also led to more reliability and performance of wind turbines. The factors discussed above make the problem of wind farm design very important, which has received attention over the past two decades [4]. The wind energy sector holds great potential for economic growth, creates jobs, and helps mitigate the adverse effects of climate change.

Understanding Wind Farm Site Selection

Selecting an optimal site for a wind farm involves evaluating multiple factors to maximize energy production while minimizing environmental and social impacts. Performance of the energy system is directly related to the geographic properties of a region, such as local resources, distance to roads, economic activities, settlement structures, and energy infrastructure [5]. Finding the appropriate site for the wind power plant's construction is an essential first step before the technical project is developed. Facility location optimization provides knowledge of modeling problems related to installations in some specific regions based on certain criteria [6, 7]. Wind farm site selection includes the study of various variables [8]. Factors affecting the environment, economy, and wind energy production must be considered [9]. Numerous studies in this field have been conducted, focusing on variables that can influence the choice of the most advantageous site for wind farm construction. It is clear from reading several research studies on wind farm placement that the primary consideration for its development is site selection. This process requires the consideration of several variables, which makes decision-making difficult, besides requiring complex modeling [10]. If all requirements are not met, this can lead to problems like high production costs, low power generation, and environmental damage [11]. Fig. (**1**) depicts the critical factors considered in the selection of wind farm sites. These include wind speed and consistency, land topography, environmental impact, grid connectivity, and proximity to populated areas.

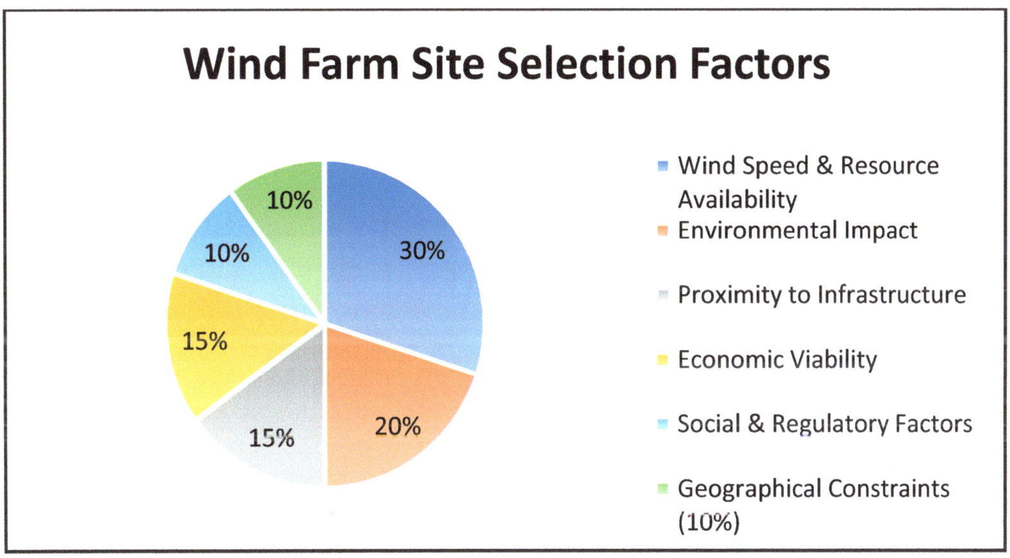

Fig. (1). Wind farm site selection factors.

Challenges in Traditional Wind Farm Site Selection Methods

The conventional procedures for determining wind turbine site suitability are subject to uncertainty arising from the arbitrary selection of criteria and model parameters by experts.

- **Data Limitations:** Historical reviews of meteorological data at the selected site and on-site measurements have been the basis by which site selection has been traditionally carried out. However, in most regions, it is observed that available wind data are sparse, not current, or do not provide the fine granularity required for an accurate assessment. This lack of data may lead to an undervaluation of a site's wind potential, resulting in low energy production and poor financial performance [11].
- **Time-Consuming Processes:** With traditional site selection methods, one visits the sites and spends months doing field surveys, environmental impact assessments, consultation with stakeholders, *etc*. Further, such delays in data acquisition, along with repeated rounds of environmental and regulatory reviews, can postpone or hamper project initiation, thereby increasing the development cost. The above-discussed constraints pose questions on the competitiveness of wind energy projects in the fast-growing energy market [8].
- **Integration Difficulties:** Integrating a wide variety of data sets, including wind patterns, topographical maps, proximity to existing infrastructure, and environmental restrictions, into a single analysis poses a large challenge. The older generation of techniques quite often lacks the modern analytical tools, resulting in gaps in the seamless integration and analysis of these data sources. Such a highly complex environment could lead to errors, omissions, wrongful conclusions, or outright exclusion of relevant information, ultimately impairing the site selection [12].

It is imperative to overcome the above obstacles for the advancement and efficiency of wind energy projects. The incorporation of high-tech Artificial Intelligence, among other technologies, promises to reduce the extent of subjective calls, improve data accuracy and efficiency, and enhance the integration of heterogeneous data sources in wind farm site selection.

The Role of AI in Wind Farm Site Selection

Artificial Intelligence (AI) is bringing a revolution in the site selection of wind farms for the mitigation of conventional challenges like data scarcity, time consumption, subjectivity, and integration. With the application of AI, improvements concerning efficiency, accuracy, and sustainability in site selection and subsequent operations can be made in the wind energy sector. Detailed Analysis:

- **Enhanced Data Analysis and Wind Resource Assessment:**
  - AI enables the processing of large, complex datasets to accurately assess wind resources, leading to optimal site selection.
  - In trash assessment, a conventional approach utilizes only sparse and usually

outdated meteorological data. In most cases, AI algorithms analyze vast amounts of historical and real-time weather data, finding patterns to facilitate accurate wind resource evaluation [13]. The conversion of such variable components thereby minimizes uncertainties in energy production forecasting and ensures that the selected sites are endowed with steady and piled-up wind efforts for efficient energy generation.

- **Optimization of Wind Farm Layouts:**
    - AI facilitates the design of wind farm layouts that maximize energy production while minimizing costs and environmental impact.
    - The spatial arrangement of turbines greatly influences a wind farm's efficiency through wake effects and land-use limitations. AI-based models simulate a variety of layout configurations of the terrain, turbine interaction, and distance from infrastructure. In accounting for these factors, AI helps determine the optimal turbine locations for maximum energy gain and minimum operating cost [14].

- **Reduction of Subjectivity in Environmental and Social Impact Assessments:**
    - AI introduces objectivity into the evaluation of environmental and social impacts associated with wind farm development.
    - Human bias may interfere with the assessment of likely impacts on wildlife, ecosystems, and vibrant communities. AI systems take a data-driven approach to analyze environmental and social parameters, thus providing unbiased information about possible adverse effects. This objectivity helps inform decisions that balance energy needs, ecological preservation, and social acceptance, thereby enabling smoother project approvals and community relations [15].

- **Integration of Diverse Datasets for Comprehensive Site Evaluation:**
    - AI excels at integrating heterogeneous datasets, offering a holistic approach to site selection.
    - The selection of a site for wind farm consideration should be an all-encompassing data set, encompassing wind patterns, topography, land use, and infrastructure availability. The manual synthesis of this information can be complicated with a high chance of error. AI platforms can funnel these datasets seamlessly to arrive at cohesive analyses indicative of optimal sites considering all related factors [16]. Thus, integration can ensure that selected sites satisfy technical, economic, and environmental criteria, thus increasing the feasibility and sustainability of the wind energy project.

- **Acceleration of the Site Selection Process:**
    - AI streamlines the site selection timeline by automating data analysis and decision-making processes.
    - A labor-intensive and time-consuming venture, traditional site selection often delays project execution. With AI tools, data analyses can be done rapidly, allowing the fastest selection of suitable sites. This speed not only helps diminish development timelines but also allows a faster variation for market needs or policy changes for wind energy's competitiveness in the renewable energy sector [17].

Combining AI with wind farm site selection would tackle former challenges and hence transform into one of the most efficient, cost-effective, and environmentally friendly wind energy projects. Fig. (**2**) illustrates the multifaceted role of AI in renewable energy systems, including predictive maintenance, energy demand forecasting, performance optimization, fault detection, and integration with smart grids. AI enhances decision-making, efficiency, and sustainability in modern energy infrastructure.

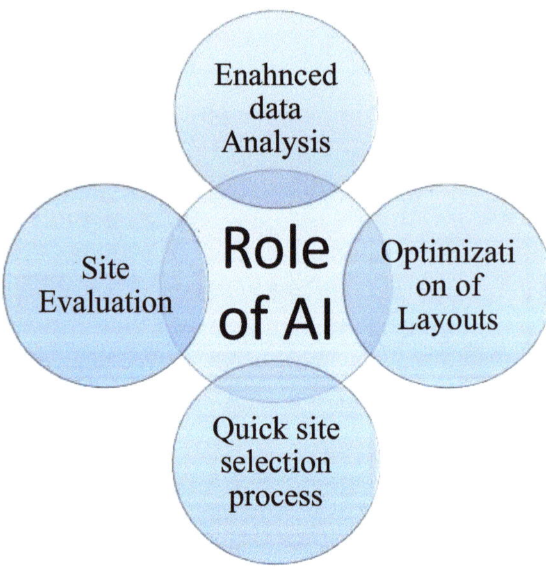

Fig. (2). Role of AI.

INTRODUCTION TO MACHINE LEARNING ALGORITHMS

The adoption of machine learning (ML) algorithms in wind site analysis has notably enhanced the precision of wind resource evaluation, optimum turbine positioning, and general efficiency in terms of energy generation from wind energy.

- **Artificial Neural Networks (ANNs)**
  - ANNs are computer models inspired by the neural networks found in the human brain. They can learn complicated patterns from input because they are made up of interconnected nodes (neurons) stacked in layers. ANNs are employed to predict wind speeds and energy output by analyzing historical weather data and identifying non-linear relationships between variables [18].

- **Support Vector Machines (SVMs)**
  - Supervised learning includes models like SVM used for classification and regression. They work by locating a hyperplane that divides the data into separate classes. In wind energy, SVMs assist in classifying suitable and unsuitable sites based on various parameters such as wind speed, topography, and environmental constraints [19].

- **Random Forest**
  - Random forest belongs to an ensemble learning method, creating many decision trees during training and averaging their predictions for regression problems or returning the class that was most voted for by the individual trees in the case of classification problems. They are used to estimate hub-height wind speeds by analyzing surface-level data, thereby aiding in the assessment of wind resources at potential sites [20].

- **Deep Learning Models**
  - Deep learning, a subset of ML, employs neural networks with multiple layers to simulate complex data patterns. Convolutional Neural Networks (CNNs) and Recurrent Neural Networks (RNNs) address very accurate wind forecasting and optimization of turbine performance using deep learning models [21].

- **Metaheuristic Algorithms**
  - Metaheuristic algorithms are high-level processes aimed at developing or selecting heuristics that provide adequate answers to optimization issues. Genetic Algorithms (GAs) and Particle Swarm Optimization (PSO) are employed to optimize the layout of wind farms, ensuring maximum energy capture and minimal wake losses [14].

- **Reinforcement Learning (RL)**
  - It means training agents to make sequences of decisions by rewarding the desired behaviors and punishing the undesired ones. Adjusting turbine settings, such as blade pitch and yaw angles, can be done by RL algorithms in response to changing wind conditions, thereby maximizing energy output

[22].

- **Clustering Algorithms**
    - Clustering brings objects together based on their similarity. Clustering algorithms, such as K-Means, are used to identify regions with similar wind patterns, which aids in regional wind resource assessment and site selection [23].

- **Recurrent Neural Networks (RNNs) and Long Short-Term Memory Networks (LSTMs)**
    - Recurrent Neural Networks or RNNs are a kind of artificial neural networks that recognize patterns in sequences of given data, such as time-series data, which is very important for wind speed and power prediction.

- **Long Short-Term Memory or LSTM networks**
    - Classic RNNs may have trouble in handling long-term dependencies due to challenges like vanishing gradients. To overcome this challenge, LSTM networks were introduced as a special kind of RNN with a particular ability to learn and memorize long sequences. Their architecture allows them to model the temporal dependencies and variability inherent in wind patterns, leading to more accurate and reliable forecasts [24]. This capability is crucial for optimizing the integration of wind energy into power grids and for effective energy trading.

- **Gaussian Process Regression (GPR)**
    - Gaussian Process Regression (GPR) is a non-parametric, probabilistic regression model that offers a versatile way to handle data uncertainty. It establishes a prior over functions and updates it with observed data to create predictions, providing not only mean estimates but also quantifying their uncertainty. In dealing with wind speed modeling and prediction at various locations and heights, especially in sparse data situations, GPR has found applications. Its probabilistic nature allows the integration of uncertainty estimates within the predictions, which is crucial for risk assessment and decision-making in wind energy projects. By offering a measure of certainty in the forecasts, GPR aids in the optimization of turbine placements and proper management of wind farm operations [25].

Fig. (**3**) presents common algorithms used for wind farm site selection, including machine learning models (*e.g.*, Random Forest, SVM), optimization techniques (*e.g.*, Genetic Algorithm, AHP), and GIS-based decision support systems. These algorithms evaluate multiple factors such as wind resources, terrain, environmental impact, and grid access to identify optimal locations.

Advantages of Artificial Intelligence

The integration of Artificial Intelligence (AI) in wind energy farms is revolutionizing the efficiency, maintenance, and management of wind power generation. AI-driven solutions are transforming the wind energy sector, making it more efficient and economically viable. Some of the advantages are listed below:

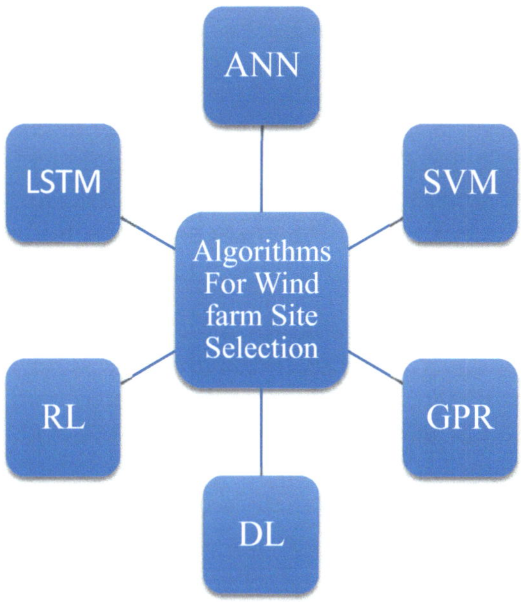

Fig. (3). Algorithms for wind farm site.

- Real-time analysis and performance monitoring of the specific turbine sites are achieved with the integration of AI into wind turbines. Such algorithmic Artificial Intelligence would automatically configure the blades and yaw angles whenever there was a shift in wind conditions and thus increase intake efficiency and enhance performance overall.
- A predictive maintenance technique based on AI algorithms analyzes large volumes of data obtained through turbine sensors. Such algorithms are capable of discovering patterns as well as anomalies, which probably signify the possibility of component failures.
- AI-enabled asset management within wind farm systems allows both monitoring and control of wind farms remotely, meaning that operators will be given centralized access to real-time data on the systems. Hence, the requirement of physical presence needs to be reduced, thus reducing the overall operational costs while improving efficiency because response time to the eventualities of

possible issues increases overall.

- Incorporating AI into wind operations could yield significant cost savings. AI could offer optimization on virtually every aspect of wind farm management, ranging from predictive maintenance to operational efficiency- to name a few examples where costs could be slashed. In fact, this could reduce operational costs overall, and those wind energy projects would become even more economically attractive by making operations more efficient by using AI.

TRADITIONAL METHODS OF WIND FARM SITE SELECTION

The complicated and critical task of finding an ideal site for a wind farm requires extensive evaluation of a host of environmental, technical, and economic considerations. Site selection by the conventional methods looks at manual data collection, statistical modeling, and environmental impact assessment in determining the feasibility of a site. The next sections outline key procedures the industry has traditionally used for wind farm site selection.

- **Wind Resource Assessments:** Wind resource assessments are the fundamental first step in identifying a suitable location for a wind farm. This involves analysis of historical meteorological data, wind speeds, and seasonal patterns to assess whether a site has wind energy potential. Such studies involve installing anemometers and using LiDAR for on-site measurements to acquire wind data over long periods, typically one to three years. The acquired data are analyzed using statistical models like the Weibull distribution for the prediction of wind behavior over time. An accurate assessment of wind resources is critical, for even minor variations in wind speed estimates could impinge on revenues and the financial feasibility of any project [26].
- **Proximity to power infrastructure**: The distance of the wind farm site from existing power sites is an important factor in reducing the losses through transmission and costs of the entire project. Ideally, wind farms are to be located close to the substation and the line of high voltage transmission, for an adequate evacuation of the power. Site selection based on conventional techniques is done by manual mapping and consideration of feasibility, which is often time-consuming and prone to errors [27].
- **Environmental Impact Assessment (EIA):** Each large-scale wind farm project shall be liable for an EIA to assess its impact on biodiversity, land use, and local ecosystems. These assessments assess noise pollution, bird and bat mortality, habitat disruption, and visual aesthetics. Traditional EIA methods consist of field surveying and stakeholder consultation with consequent predictive modeling of expected ecological consequences [26].
- **Land Use Policies and Zoning Regulations:** Wind farm development is limited by land use policies and zoning regulations that define zones suitable for the

development of renewable energy projects. The traditional selection of sites would involve checking land use maps and interaction with stakeholders to address land ownership claims. A local opposition to wind farms is often articulated on the grounds of property devaluation and adverse changes in scenic value, hence impacting the feasibility of project implementation. All these could be mitigated through clear communication, community involvement, and policy incentives that would encourage landowners to lease land for wind energy development [28].

- **Economic Feasibility Assessment:** Economic feasibility assessments are the yardstick for measuring the financial viability of the wind farm project. Traditional methods include cost-benefit analysis, levelized cost of electricity (LCOE) calculations, and evaluation of investment risk. This assessment takes into account capital costs, operation and maintenance costs, revenue forecasts, and relevant subsidy schemes applicable to a project. Economic-modelling techniques like Monte Carlo simulations are generally employed in the understanding of the overall effect on project profitability of fluctuation in wind speeds, energy prices, and policy changes [29].
- **Weather and climatic conditions:** Weather and climatic conditions are the major determinants of the long-term prospects of a wind farm. Traditionally, historical accounts of weather and variations in weather variables like temperature and air pressure are analyzed to assess behavior in relation to wind. These assessments yield information about seasonal variations with extremes like hurricanes, storms, and frost that might affect turbine operation and longevity. Engineers rely on information about wind speed and direction from a meteorological tower or alternative uses of remote sensing technologies such as LiDAR and SODAR at various elevations [30].
- **Structural integrity:** Structural integrity depends on the foundation of the soil and geotechnical conditions. Traditional site selection methods depend on conducting geotechnical surveys that analyze soils, bedrock depth, and groundwater levels. Engineers will determine soil load-bearing capacity before discussing foundation design, and thus, the stability of turbines is ensured for the entire wind-load force. Traditionally, manual drilling and soil samples are time-consuming and costly [26].
- **Accessibility of the site of wind farm construction:** Accessibility of the site of wind farm construction is critical for the project's feasibility, as well-established infrastructure is a precondition before moving large turbine components. The traditional methods of site selection involve the assessment of the existing network, including roads, bridges, and transport routes, concerning the extent to which turbine blades, nacelles, and towers can be shipped without suffering great logistical impediments. Steep slopes, river crossings, and other such conditions related to soft ground may further affect turbine installation. Most of

the logistics planning is done based on manual route assessments and feasibility studies, which can often lead to errors [26].
- **The policy and judicial framework:** It includes incentives from the government, which determine the feasibility of bringing a site to be used for wind farm development. Site selection usually involved the screening of long and torturous permitting procedures; obtaining environmental clearances; and following local, national, and international regulations.

Challenges in Traditional Methods of Wind Farm Site Selection

- Wind resource assessment is a prolonged process because data must be collected over long periods, ranging from 1 to 3 years, during which the effects associated with climate change may not have been fully taken into account, and thus cause inaccuracies in long-run energy yield functions.
- The distance between sources of supply is crucial in remote areas, where large investments must be made to develop infrastructure, while grid capacity constraints could render projects unfeasible.
- Long periods are involved in the Environmental Impact Assessment (EIA) process for approval, and indicate uncertainties regarding the long-term adverse effects of wind farms on birds and bats.
- Land use and zoning codes can be rather complex and may delay project implementation, since one major barrier to implementation is community opposition.
- Bad predictions relating to energy yield can skew the economics of the feasibility study. All these factors are going to affect project finance, and if they get even more skewed, then there will arise concerns about risk perception for investment, arising out of general economic factors like changes in policies and inflation.
- Advertisement and climate conditions may be very difficult to accurately predict due to scarce historical climate data, and extreme weather conditions may affect the endurance of turbines and the costs of maintenance.
- Geotechnical and stability investigations of soil are lengthy and costly, and soil conditions varying across a site require detailed site-specific assessments.
- Access and transport logistics can be highly challenging where access is through untraversed terrain or hilly terrains, and may add enormous costs and logistical problems in ferrying turbine components. Arable Productions: A Thrust on Changes in Municipal Policy.
- Social acceptance and the issue of community acceptance are no longer problems; the opposition at the local level from residents is capable of delaying and stopping wind farm projects, irrespective of the engagement efforts that will have been put in place through surveys and public meetings.
- The prevailing policy and regulatory framework undergoes change; hence, delay

in permitting and approvals becomes a crucial factor contributing to an increase in project duration, impacting overall feasibility. Table **1** summarizes traditional methods used for wind farm site selection, including approaches such as expert judgment, GIS mapping, and multi-criteria decision analysis (MCDA). It outlines their core features, advantages such as simplicity and low computational demand, and limitations like subjectivity and limited scalability.

Table 1. Summarizes traditional methods of wind farm site selection, including their features, advantages, and limitations.

Method	Description	Key Features	Advantages	Limitations
Wind Resource Assessment	Evaluates wind speed, direction, and consistency to determine energy potential.	Uses meteorological towers, LiDAR, and historical weather data.	Helps estimate energy yield and optimize turbine placement.	Time-consuming, requires long-term data collection.
Proximity to Power Infrastructure	Determines distance to substations and transmission lines to minimize energy losses and costs.	Uses GIS mapping and feasibility studies to assess grid connectivity	Reduces transmission losses and ensures efficient energy distribution.	Remote locations require costly infrastructure development.
Environmental Impact Assessment (EIA)	Evaluates potential effects on wildlife, land, and ecosystems.	Includes noise, bird and bat mortality, habitat disruption.	Ensures compliance with environmental regulations.	Lengthy approval process, uncertainties in long-term environmental effects.
Economic Feasibility Analysis	Assesses costs, revenue potential, and investment risks.	Uses Levelized Cost of Electricity (LCOE), and Monte Carlo simulations.	Helps investors make informed financial decisions.	Inaccurate energy yield predictions can impact financial viability.
Meteorological and Climatic Conditions	Studies long-term climate trends and extreme weather risks.	Analyzes seasonal variations.	Ensures turbines are designed to withstand weather fluctuations.	Limited historical climate data can lead to inaccurate forecasts.
Policy and Regulatory Frameworks	Navigates legal and governmental approvals.	Involves securing permits, tax incentives etc	Ensures legal operations	Changing policies can slow project timelines.

OPTIMIZING WIND FARM SITE SELECTION WITH AI

Advanced AI Algorithms for Selecting Optimal Wind Farm Sites

The adoption of AI algorithms in wind site analysis has notably enhanced the precision of wind resource evaluation, optimum turbine positioning, and general efficiency in terms of energy generation from wind energy.

- **Artificial Neural Networks (ANNs):** They have a very broad application field in wind resource assessment. An artificial neural network consists of layers of interconnected neurons that process input into output *via* computations. In wind farm site selection, ANNs study the meteorological data over a period of time in terms of wind speed, wind direction, and other climate parameters to determine potential areas for wind energy production. The most prominent advantage of ANNs is the fact that they are adaptive and, therefore, keep improving the accuracy of forecasts as they continue to learn from new input data. They can also be incorporated into GIS-based frameworks to enhance site selection by assessing the distribution of wind resources and other environmental constraints [31 - 33].
- **Support Vector Machines (SVM):** Support Vector Machines are solid supervised learning algorithms that can widely engage in the study of prediction over a variety of objectives, such as distinguishing sites for wind farms. SVMs create optimal hyperplanes for separating data from different categories, especially useful in the classification techniques that analyze wind patterns, terrain characteristics, and land-use constraints. For wind site classification, SVM models classify sites based on the criteria of site: wind intensity, geographic accessibility, and impact assessment on environmental disposal. Specific benefits of SVM include its capability of dealing with high-dimensional data, preventing overfitting, and ensuring strong models for the selection of sites. Researchers have demonstrated that the SVM models in combination with GIS datasets vastly improve the site classification accuracy as well as efficiency for decision making [34].
- **Random Forest (RF) Algorithm:** Random Forest (RF) is an ensemble learning method that employs multiple decision trees to evaluate various environmental, geographic, and economic parameters, enhancing the process of wind farm site selection. The RF model has a special significance while planning wind energy as it handles missing or inconsistent data, which is the inherent fallacy related to almost every wind resource assessment. The potential wind farm sites are ranked in terms of overall suitability based on historical wind speed, grid-infrastructure proximity, and environmental impact consideration. The inbuilt variable selection process in RF leads to prioritizing some key variables for more precise decision-making. Also, RF has been proven effective in GIS-integrated models analyzing the spatial patterns of wind energy potentials and optimizing turbine placement strategies. Empirical studies indicate that RF-based models offer a relatively high predictive accuracy in wind farm planning, hence preferring them in large-scale assessments [35].
- **Genetic Algorithms (GA):** Genetic Algorithms (GA) are optimization techniques inspired by natural selection principles, widely applied in wind farm planning to achieve multi-objective optimization. GAs work by simulating

evolutionary processes such as selection, crossover, and mutation to find the best combination of wind turbine locations, orientations, and capacities. This way, one remains within an optimum between the economic, environmental, and technical constraints. An advantage of using GA is that one can efficiently search along vast solution spaces to identify the placement of the turbines that yield the most wind energy generation while optimizing the environmental effect. Models based on GA have been applied to improve layouts in wind farms, showing increased energy efficiency and cost-effectiveness than the traditional planning methods. Studies show that GA optimizes the spatial distribution and wake effects and thus increases wind power utilization efficiency, which makes it a vital technology for sustainable wind farm development [36].

- **Fuzzy Logic Systems:** Fuzzy Logic Systems (FLS) are computational models for uncertainty and imprecise data handling in wind farm site selection. Fuzzy models basing site suitability on the less-known variables of wind variability, land characteristics, regulatory policies, and environmental limits can realistically rank potential sites on the degrees of their assets rather than just without absolute classifications. GIS integrated with fuzzy logic frameworks has been used in improving site selection criteria, thus refining its determination. The effectiveness of fuzzy logic in managing uncertainty in wind resource assessments has been emphasized by the researchers, bringing about a more reliable outcome in terms of site selection [37].
- **Deep Learning (DL) Techniques:** The latest methods for wind farm site selection utilize Deep Learning (DL) techniques, mainly Convolutional Neural Networks (CNNs), as the preprocessing step prior to the human checking step. CNNs handle large-scale geospatial and meteorological data to identify patterns essential for wind potential assessment. Deep learning models have long been appreciated for wind speed forecasting and energy production estimation, which typically result in better-informed sites for the planned wind farms. Studies also indicate that geospatial analysis with CNNs will play a major role over time in detecting wind corridors and land suitability patterns to optimize identified wind farm locations. And as the study of DL continues, both GIS and remote sensing techniques are anticipated to be integrated further to promote wind energy planning [38].

This is how Artificial Intelligence (AI) algorithms enhance the wind farm site selection by improving the accuracy, increasing the efficiency, and enhancing the decision-making process. The application of these AI methods yields a more improved wind farm site selection with its maximum energy production, minimum environmental damage and cost inefficiencies [37, 38]. Table **2** summarizes the key features of Artificial Intelligence (AI) algorithms employed in wind farm site selection. These include the ability to handle large and complex datasets, learn patterns from historical data, support multi-factor decision-making, improve prediction accuracy, and adapt to changing environmental conditions. Common AI techniques include machine learning, deep learning, and hybrid optimization models.

Table 2. Summarizes the key features of the AI algorithms used for wind farm site selection.

Algorithm	Key Features	Advantages	Limitations
Artificial Neural Networks (ANNs)	ANNs study the meteorological data over a period of time to determine potential areas for wind energy production.	High accuracy in wind energy prediction, adaptable to new data.	Requires large datasets, computationally expensive.
Support Vector Machines (SVMs)	Classifies potential sites based on wind speed, land use, and terrain constraints.	Effective for high-dimensional data, it prevents overfitting.	Computationally intensive for large datasets.
Random Forest (RF)	Uses multiple decision trees to analyze environmental, geographic, and economic factors.	Handles missing data well, ranks site suitability based on multiple parameters.	It can be slower for large datasets, prone to overfitting if not tuned properly.
Genetic Algorithms (GAs)	Simulates natural selection to optimize turbine layout and site selection.	Finds optimal solutions for complex problems and balances multiple constraints.	Requires careful parameter tuning; results may vary per execution.
Fuzzy Logic Systems (FLS)	Handles uncertainty and imprecise data, integrates expert knowledge into decision-making.	Flexible, effective in uncertain conditions, enhances GIS-based selection.	Requires expert-defined rules; performance depends on input quality.
Deep Learning (DL) (CNNs)	Analyzes satellite imagery and climate data to predict wind energy potential.	High accuracy in detecting wind patterns.	Computationally intensive, needs large datasets for training.

Integration of AI with Data Sources

Wind farm site selection relies on diverse and highly qualified data sources to inform precise decision-making. The main sources of data include satellite imagery and remote sensing, meteorological and climate data, and GIS data. Fig. (**4**) illustrates the integration of Artificial Intelligence (AI) with various data sources used in wind farm site selection. These data sources include meteorological data (*e.g.*, wind speed, air density), geographical information system (GIS) data (*e.g.*, terrain, land use), environmental constraints, and grid infrastructure. AI algorithms process and analyse this multi-source data to generate accurate, data-driven site recommendations.

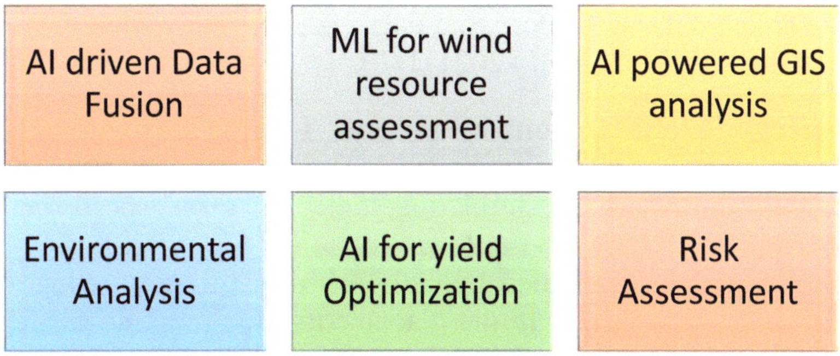

Fig. (4). Integration of AI with data sources.

- **AI-Driven Data Fusion for Site Selection:** Linking satellite imagery, meteorological information, and GIS data into a single hub is possible with AI. Synthesis techniques reveal relevant information from heterogeneous formats and resolutions, increasing accuracy in site selection. The datasets are analyzed through machine-learning models such as Random Forests or Neural Networks to find possible identical patterns that might be unrecognizable in isolated views. By merging remote sensing with meteorological data, AI refines wind prospect estimates to optimize the efficiency of wind farms (Zhang *et al.*, 2023) [38].
- **Machine Learning for Predictive Wind Resource Assessment:** Through its learning of historical meteorological patterns, AI algorithms predict wind speed and, therefore, energy output from the wind. Among the preferred methods for wind speed predictions from past climate records are Long Short-Term Memory (LSTM) networks and Support Vector Machines (SVM). Their AI-driven weather models make better predictions of the variations in wind speed compared to the traditionally statistical models, with their processing of data that includes satellite-derived cloud movement images, atmospheric pressures, and temperature. This long-term predictive ability allows the developer to assess long-term feasibility before investing in a wind farm (Liu *et al.*, 2022) [39].
- **AI-Powered GIS Analysis for Optimal Site Identification:** Through Multi-Criteria Decision Analysis based on GIS, the AI part presents potential site assessment automation based on wind availability, land restrictions, and grid accessibility. Deep learning models, such as Convolutional Neural Networks (CNNs), learn to analyze these spatial datasets and detect such features on land as forests, water bodies, and urban areas to minimize intervention with the environment.

- **Automation of Terrain and Environmental Analysis:** Emerging technologies like the AI paradigm process remote sensing data for an insight into terrain attributes concerning elevation, surface roughness, and land cover. Ecologically sensitive areas can be avoided by adaptive Artificial Intelligence that integrates wildlife movement data with geospatial datasets to forecast wind turbine impacts on local biodiversity (Riahi *et al.*, 2021) [40].

- **AI for Energy Yield Optimization and Risk Assessment:** AI allows predictions on the energy yields within a particular site in different meteorological conditions by simulating the wind farm's performance under varying conditions. The total virtualized model of the wind farm, which tests different scenarios by AI-enabled Digital Twin Technology, reduces the wake effect losses after optimizing turbine placements. Such risk assessment models based on AI also identify potential hazards even before they happen, such as with extreme weather events or infrastructure failures (Wang *et al.*, 2023) [41].

AI-Driven Process for Wind Farm Site Selection

The right site to be chosen for a wind farm is vital for optimizing energy production as well as environmental and economic feasibility. Such site selection has improved on AI techniques through integrating very large datasets, using machine learning models for wind resource analysis, and optimizing turbine placement. The following document presents the process of wind farm site selection through AI in a structured form.

- **Data Collection and Preprocessing**

AI plays an important role in site selection by preselecting and fine-tuning different datasets.

Data Sources:

- Satellite Imagery: Satellite Imagery is providing high-resolution images from NASA's Landsat and ESA's Sentinel for topographical and land-use information (Perveen *et al.*, 2021).
- Meteorological Data: Historical and real-time wind data from such sources as ERA5 (ECMWF) and NOAA are necessary for an analysis (Liu *et al.*, 2022).
- Geographic Information System (GIS) Mapping: GIS maps are used to examine terrain features, land cover types, and environmental parameters (Riahi *et al.*, 2021).
- Infrastructure Proximity: It reduces logistics costs when data about those existing roads, transmission grids, and substations is available (Zhang *et al.*, 2023).

Preprocessing Techniques:

- Noise Reduction: AI filters erroneous and irrelevant data.
- Data Standardization: Ensures uniformity across datasets.
- Missing Value Imputation: Methods like K-Nearest Neighbors (KNN) fill data gaps.
- Dimensionality Reduction: Principal Component Analysis (PCA) extracts critical features, reducing computational load. As shown in Fig. (**5**), the preprocessing pipeline includes normalization, feature selection, and model training using supervised learning techniques.

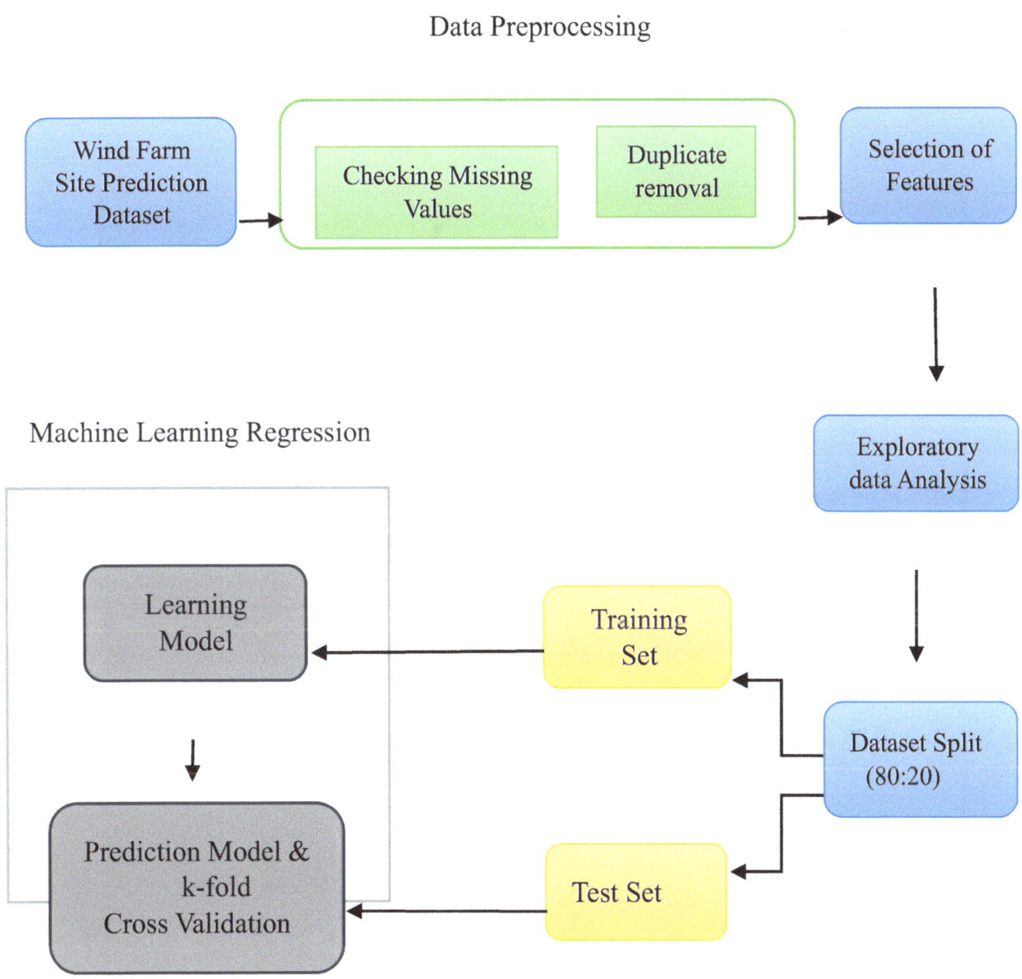

Fig. (5). Preprocessing and model training.

WIND RESOURCE ASSESSMENT USING AI

The optimal energy generation takes place if wind resources are accurately assessed.

AI Models Used:

- Recurrent Neural Networks (RNNs): This examines past wind activity to predict the smaller and longer patterns of wind (Wang *et al.*, 2023).
- Long Short-Term Memory (LSTM) Networks: This captures long-term dependencies in the wind speeds and directions and augments future prediction accuracy (Liu *et al.*, 2022).
- Gaussian Process Regression (GPR): Estimates wind power density in data-sparse regions while quantifying uncertainty (Zhang *et al.*, 2023) [42].

Data Integration:

- AI models combine satellite observations, meteorological data, and reanalysis datasets (*e.g.*, ERA5) to improve wind predictions.
- AI-generated wind maps assess wind consistency and turbulence.

OPTIMIZATION OF WIND TURBINE PLACEMENT

Proper turbine placement enhances efficiency and reduces wake effects.

AI-Based Optimization Techniques:

- Genetic Algorithms (GA): AI-driven evolutionary models optimize turbine layout by simulating multiple configurations (Perveen *et al.*, 2021).
- Computational Fluid Dynamics (CFD) Models: Predict airflow patterns to minimize energy loss (Riahi *et al.*, 2021).
- Reinforcement Learning: AI models refine turbine placement based on real-world data (Zhang *et al.*, 2023).

Key Considerations:

- Turbine Spacing: Prevents wake interference.
- Wind Flow Analysis: Ensures optimal turbine alignment.
- Topographical Features: AI evaluates terrain impact on wind flow [43].

ENVIRONMENTAL AND SOCIOECONOMIC CONSIDERATIONS

AI helps in assessing the environmental and social impacts of wind farms. The AI-driven process for wind farm site selection is illustrated in Fig. (**6**), which includes data acquisition, preprocessing, model training, evaluation, and site ranking.

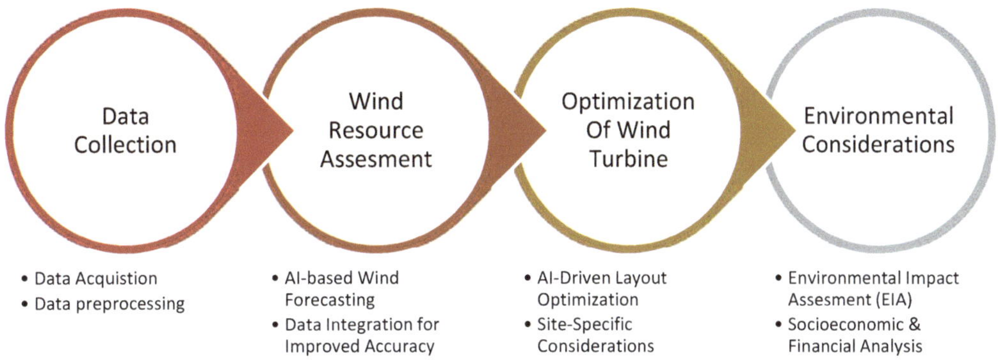

Fig. (6). AI-Driven Process for wind farm site selection..

Environmental Impact Assessment (EIA):

- Biodiversity Mapping: AI-powered remote sensing identifies protected habitats.
- Noise and Visual Pollution: AI models simulate turbine noise and aesthetic impacts (Riahi *et al*., 2021).
- Carbon Footprint Estimation: AI calculates emission reductions from wind energy.

Socioeconomic Considerations:

- Land Use Compatibility: AI assesses agricultural and residential impacts.
- Economic Feasibility: AI models forecast financial returns, maintenance costs, and incentives.
- Public Sentiment Analysis: AI evaluates community feedback to improve project acceptance.

ADVANTAGES OF WIND FARM SITE SELECTION WITH AI OVER TRADITIONAL METHODS

AI models can increase accuracy, efficiency, time, and cost in the selection and evaluation of wind energy sites, energy yield prediction, environmental impact assessment, data-based decision-making, public and stakeholder engagement, and adaptation to change. It uses large data sets, reduces feasibility study expenditure, and updates models continuously for climate change issues.

- **Enhanced Accuracy and Efficiency:** AI intelligent models utilize an extensive amount of data ranging from satellite images, meteorological information, and historical wind patterns to make more accurate site selection decisions.
- **Reduced Time and Cost:** While traditionally involving years of data gathering, AI models can, within a short time, process huge quantities of data, thus drastically reducing feasibility study costs while fast-tracking the decision-making process.
- **Optimized Energy Yield Prediction:** Most of the traditional methodologies have a hard time because of the inaccuracies that have resulted from modifying climate patterns, while AI improves its long-term forecasting through constant updating of its models using the latest data.
- **Expanded Environmental Impact Assessment:** AI also extends ecological data analysis, yielding insights into the migration patterns of wildlife and potential biological threats compared to traditional EIA.
- **Data-Driven Decision Making:** Unlike traditional methods that would have static datasets that lose their worth with time, AI combines real-time information on weather, topography, and constraints in planning to locate optimal sites for wind farms.
- **Enhanced Public and Stakeholder Engagement:** People and stakeholders do not usually resist AI-simulated models and presentations when demonstrated to them, unlike traditional public consultations, where resistance might be precipitated as communities question where the actual, back-up evidence lies.
- **Adaptation to Change:** AI keeps learning with the new climate data, modifying the models. Conventional methods do not seem to be sufficient or capable of addressing the changes in an evolutionary climate devolving, which leads to lower resilience in planning wind farms. Table **3** summarizes the key features of the AI algorithms employed for wind farm site selection, highlighting differences in model complexity, training time, and accuracy.

Table 3. Here is a summary of the key features of the AI algorithms used for wind farm site selection.

Criteria	AI-Based Site Selection	Traditional Site Selection
Speed	Very fast analysis of large data within days.	Takes years for wind resource assessment and feasibility studies.
Accuracy	Machine learning improves precision in energy yield prediction.	Limited by static datasets and long-term climate uncertainties.

(Table 3) cont.....

Criteria	AI-Based Site Selection	Traditional Site Selection
Cost	Lower feasibility study costs due to automation.	High costs due to field surveys, environmental studies, and expert consultations.
Environmental Impact Analysis	AI detects risks using satellite and wildlife tracking data.	Manual surveys and assessments may overlook critical ecological factors.
Stakeholder Engagement	Uses AI-powered simulations to improve communication.	Relies on town hall meetings, which may face resistance and delays.
Climate Adaptability	Dynamic adaptation to changing climate patterns.	Limited ability to adjust predictions based on new climate data.
Geotechnical Analysis	AI-driven remote sensing enhances soil and terrain analysis.	Expensive and time-consuming field-based assessments.

ENVIRONMENTAL AND ECONOMIC IMPACTS OF AI-OPTIMIZED SITE SELECTION

AI-based site selection for wind farms eliminates land use conflicts, lowers carbon footprint, protects animals and ecosystems, increases cost-effectiveness and investment returns, improves energy production efficiency, and adjusts to climate change.

- **Minimizing Land Use Conflicts:** AI site selection assures siting wind farms where there are going to be minimal land-use conflicts. AI-based models assess multiple criteria, such as soil composition, vegetation cover, and settlement patterns, which optimize the location choices, thus minimizing legal and environmental issues (Perveen *et al.*, 2021)
- **Reducing Carbon Footprint**: Optimised site selection using AI has also contributed to reducing the carbon footprint of the wind power farm. The energy generated per turbine is maximized with the least number of turbines located in the areas with the most wind energy potential, resulting in a reduction in construction materials and emissions. Studies indicate that AI optimization can bring about 10-20% improvement in power efficiency, which directly results in greenhouse gas emissions associated with electricity generation (Riahi *et al.*, 2021).
- **Protecting Wildlife and Ecosystems**: Generally, the siting of a traditional wind farm produces a detrimental alteration in the structure of habitats for the local wildlife populations, particularly birds and bats. AI incorporates remote sensing and environmental impact assessments in its suggestions for alternate sites bearing reduced risk to wildlife. Machine-learning training, which combines past

ecological impacts, predicts the scheduled potential disruption and advises adaptive turbine shutdowns when migration peaks (Liu *et al.*, 2022).

- **Improving Cost-Effectiveness and Investment Returns**: AI-driven site selection defines the economic potential of wind farms by determining the locations with the maximum return on investment (ROI). Predictive analytics models simulate how data about long-term wind conditions, energy output, and maintenance costs allow developers to make informed decisions. Combining this with uncertainty improvements and better optimization of resource allocation should translate to a reduction in financial risk for the investor. Up to 30% higher profitability can be realized by optimizing the location of the wind farms compared to the traditional site selection methods (Zhang *et al.*, 2023).
- **Enhancing Energy Production Efficiency**: AI would optimize the space field layout of wind farms such that maximum energy is harnessed with minimum infrastructure. State-of-the-art algorithms analyze the topographic data, the wind flow directions, and the turbine systems wake effects to place the turbines at positions where they would work at their best. AI optimization approaches can therefore contribute an actual annual energy production gain of between 5 to 15%, allowing wind farms to roll out economically and with a reduced environmental impact (Wang *et al.*, 2022).
- **Reducing Construction and Maintenance Costs**: The predictive models of AI that apply statistical learning minimize construction costs by selecting the ideal sites with less topography interference. AI-enabled predictive maintenance strategies use predictive analytics to signify a potential failure before it happens, reducing downtime and repair costs.
- **Adapting to Climate Change and Extreme Weather Events:** AI will interpret historical and real-time meteorological data to predict possible variations in wind speed due to extremely different weather conditions, as well as seasonal shifts in the wind patterns to choose areas that can be productive for decades to come, thus minimizing risks associated with climate change. Thus, AI brings climate-smart adaptation strategies to make wind farms more resilient and sustainable investments (Xu *et al.*, 2023) [44].

APPLICATIONS OF AI IN WIND ENERGY: OPTIMIZATION, EFFICIENCY, AND SUSTAINABILITY

AI is changing the future of wind energy, especially for optimizing wind farm site selection. It uses advanced algorithms to enhance performance, sustainability, and efficiency within a wind energy system. Key applications of AI in this domain are described as follows:

- **Improved Wind Resource Assessme**nt: Proper wind resource assessment is a basis for wind farm deployment. AI helps to analyze large amounts of data for a precise wind pattern prediction using machine learning models from historical weather data, topography, atmospheric conditions, and wind speeds and directions. With such a forecast, developers can locate turbines at the most optimal positions to maximize energy capture. AI-driven wind resource assessments result in reduced uncertainties in energy production estimates and increased reliability and efficiency in wind farm operations.

- **Optimal Turbine Placement by means of Clustering Algorithms:** A wind farm energy outcropping and operational efficiency are all best supported by placing turbines strategically. AI employs clustering algorithms for grouping areas with similar wind characteristics by analyzing wind data and terrain features. This informs the optimum layout of the turbines, limiting wake effects by updrafts and increasing exposure to the more consistent wind flows. Optimized turbine placement increases the overall energy yield of the wind farm and reduces mechanical stress on turbines, thereby extending their operational lifespan.
- **Integration with Energy Grids:** Effective integration of wind energy with the existing power grid was needed for matching demand and supply, so AI would carry out efficient scheduling and dispatch for wind production. Advanced algorithms for demand-side management take care of the unpredictable nature of wind energy production, ensuring reliability and stability for supply. AI-enabled grid integration is, therefore, strengthening the reliability of renewable energy sources, enabling better penetration of the rest of the wind power, and enhancing the transition towards a sustainable energy system.
- **Real-Time Performance Optimization:** Continuous resource management and modifications in the operation of turbines are vital to achieving optimal energy production. One instant adjustment of operational parameters can be done by real-time data analysis of wind conditions and turbine functionality to estimate the dynamic changes of wind resources. Instant AI interventions achieve greater output of energy, lesser wear of equipment parts, and higher overall economic profitability of wind projects.
- **AI-Driven Wind Energy Forecasting:** Accurate forecasting of wind energy production is crucial for better grid integration and a mechanism for operational planning. AI models utilizing historical weather data with rcal-time meteorological requirements enable forecasting energy production for both short- and long-term time scales. Implementing AI forecasts improves the reliability of wind energy, reduces operational costs, and stabilizes the power grid.
- **AI-Based Fault Detection and Predictive Maintenance:** Anticipating faults beforehand in wind turbines is fundamentally necessary to avert unscheduled failures, along with a huge cost involved in downtimes. AI-based algorithms would analyze the effect of data taken from turbine sensors and recognize

patterns that indicate possible wear or failure. Early pattern recognition allows maintenance to be carried out beforehand and not dependent on issue escalation severity. Such conditioning of turbines increases the reliability, life, and continued reduction in maintenance costs as well as energy production losses.

- **AI in Wind Turbine Health Monitoring:** For turbines to operate perfectly and avoid failures, monitoring is necessary: turbines should be monitored in real time to determine their health constantly. AI assesses the information derived from sensors to spot anomalies and predict when parts will fail.
- **AI-Assisted Design and Optimization of Wind Farms:** Wind farm design is complex, requiring determining the optimal configuration of turbines concerning where they are placed, what type of turbine design to use, and finally, how to improve energy production as much as possible. Different design scenarios would be simulated by using AI algorithms in optimizing turbine configuration for defining the terrain, wind patterns, and environmental impact. AI-assisted design leads to higher energy outputs coupled with reduced environmental footprints, making projects more cost-effective to develop.

As these technologies develop, the applicability of AI in wind energy is expected to grow and yield higher efficiencies and more innovations in the sector. Fig. (**7**) illustrates various applications of AI, including predictive analytics, optimization in renewable energy systems, and intelligent decision support for site selection.

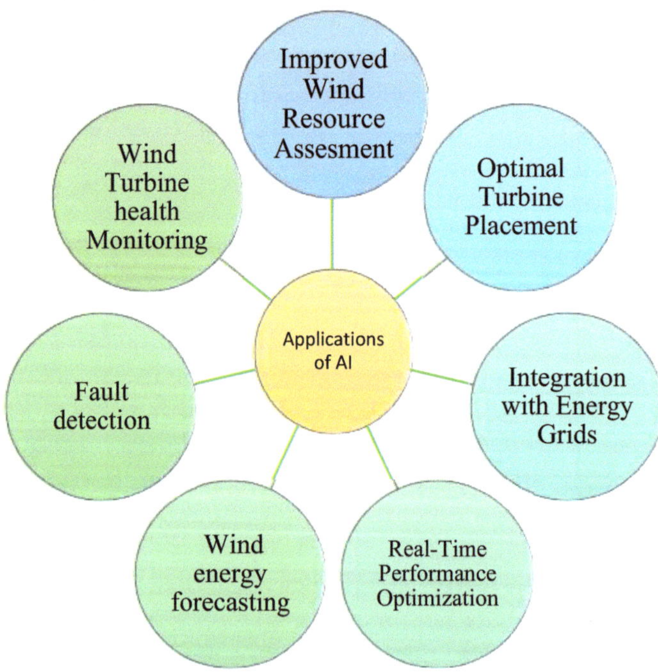

Fig. (7). Applications of AI.

AI CHALLENGES AND FUTURE DIRECTIONS IN OPTIMIZING WIND FARM SITE SELECTION WITH AI

AI is in a position to be a big decision-maker with regard to optimizing site selection for wind farms; however, challenges remain on several fronts. There is a scarcity and poor quality of data that hinders accurate prediction by the AI model, which may require covering datasets that include reliable and comprehensive information and which many regions lack [45]. Most of the models developed using deep learning still require a significant amount of computational power to give optimal results on the web, which itself might pose a big drawback for a major portion of the wind energy market. Some of the challenges are listed below:

- **Data Shortage and Quality Issues**

The AI model site selection for wind farms is directly dependent on high-quality meteorological, geographical, and environmental data. Additionally, inconsistencies in data collection methods and missing values can lead to unreliable site assessments [46].

- **Uncertainty in Long-Term Wind Forecasting**

Long-term wind forecasting is a vital task and is notoriously difficult because of the chaotic nature of weather systems. The factors, such as abrupt changes in climatic conditions in the form of temperature changes and extremes of weather events, add uncertainty for wind energy production and impairment of forecasting capacities.

- **Computational Complexity and High Processing Requirements**

AI-based site selection models, for example, those that are deep learning or genetic algorithms, require a large number of computational resources. Training deep learning models for wind energy assessment involves processing large-scale spatial and temporal data, leading to high computational costs [47].

- **Lack of Clarity in AI Models**

AI models like deep learning techniques operate as "black boxes," which makes it difficult to interpret their decision-making processes. The lack of transparency in AI-driven wind site selection can hinder regulatory approvals and stakeholder trust [48].

The AI domain is crucial for optimizing wind energy developments by improving efficiency and accelerating the global shift toward renewable energy. To address the challenges in AI-driven wind farm site selection, several future directions are being explored:

- **Enhanced Data Collection and Open-Source Datasets**

The future research should look at the standardization of wind resource databases and provide them as open-access directories to support AI prediction. Combined efforts from improved satellite imaging data acquisition, IoT-based weather sensors, and real-time data-sharing platforms can easily address issues related to data scarcity.

- **Hybrid AI Models for Improved Forecasting**

By using AI with physics-based weather models, the precision of wind power forecasts can be enhanced. Hybrid models that combine machine learning with numerical weather prediction (NWP) methods can give better results in the long term.

- **Cloud-Based AI and Edge Computing for Real-Time Analysis**

The use of cloud computing and edge AI can significantly reduce computational burdens by processing data closer to wind farm locations. This approach enables real-time analysis while reducing latency and energy consumption in AI model execution [49].

- **Explainable AI (XAI) for Better Transparency**

The development of Explainable AI (XAI) algorithms will provide meaningful information regarding selected site models. This transparency would help policymakers and energy sector players develop a better understanding of why a given site is suitable for wind farm placement.

CONCLUSION

The inclusion of Artificial Intelligence in site selection for wind farms has significantly improved efficiency and accuracy in identifying suitable sites for wind energy projects. The AI techniques adopted include Artificial Neural Networks, Support Vector Machines, Random Forest algorithms, Genetic Algorithms, Fuzzy Logic Systems, and Deep Learning, which have all contributed to sufficiently assessing the potential sites bearing in mind their ability to handle very complex datasets and even uncertainties defined by environmental and socio-economic features alongside meeting technical and economic criteria for site selection as well as contributions toward environmental sustainability goals.

Furthermore, AI is not limited to site selection but also extends its application to optimizing wind farm layouts and their operations.

All in all, applying Artificial Intelligence in the site selection and operational optimization of wind farms marks a significant advancement in the renewable energy sector. The capability of AI in processing millions of complex datasets will significantly empower stakeholders to make sound decisions and also enhance the sustainable and economically viable growth of wind energy infrastructure.

REFERENCES

[1] J. Yang, and B. Chen, "Integrated evaluation of embodied energy, greenhouse gas emission and economic performance of a typical wind farm in China", *Renew. Sustain. Energy Rev.,* vol. 27, pp. 559-568, 2013.
[http://dx.doi.org/10.1016/j.rser.2013.07.024]

[2] S. Bakhtiar, J. R. T. Wijaya, and N. Hidayah, "Artificial Intelligence in Accounting: A Systematic Literature Review and Bibliometric Analysis of Opportunities, Challenges, and Future Directions," *BISECER (Bus. Econ. Entrepreneurship)*, vol. 9, no. 1, pp. 1–22, Feb. 2026.

[3] Y. Ding, C. Shao, J. Yan, Y. Song, C. Zhang, and C. Guo, "Economical flexibility options for integrating fluctuating wind energy in power systems: The case of China", *Appl. Energy,* vol. 228, pp. 426-436, 2018.
[http://dx.doi.org/10.1016/j.apenergy.2018.06.066]

[4] I. Ul Haq, I. U. Hassan, and H. A. Shah, "Machine Learning Techniques for Result Prediction of One Day International (ODI) Cricket Match", In: *Proceedings of the 2023 IEEE 8th International Conference on Convergence Systems and Innovative Technologies (I2CT)*, 2023.
[http://dx.doi.org/10.1109/I2CT57861.2023.10126241]

[5] I. U. Haq, A. H. Rather, and G. Kaur, "A Comparative Analysis of Machine Learning Algorithms for the Early Prediction of Cardiovascular Disease", In: *Proceedings of the 2023 2nd International Conference on Edge Computing and Applications (ICECAA)*, 2023, pp. 987-993.
[http://dx.doi.org/10.1109/ICECAA58104.2023.10212214]

[6] A. Mostafaeipour, S. Sadeghi, M. Jahangiri, O. Nematollahi, and A. Rezaeian Sabbagh, "Investigation of accurate location planning for wind farm establishment: a case study", *J. Eng. Technol.*. vol. 18, no. 4, pp. 821–845, 2020.
[http://dx.doi.org/10.1108/JEDT-08-2019-0208]

[7] G. Rediske, J.C.M. Siluk, L. Michels, P.D. Rigo, C.B. Rosa, and G. Cugler, "Multi-criteria decision-making model for assessment of large photovoltaic farms in Brazil", *Energy,* vol. 197, p. 117167, 2020.
[http://dx.doi.org/10.1016/j.energy.2020.117167]

[8] A.H. Rather, and I.U. Haq, "Intelligent Framework for Early Prediction of Diabetic Retinopathy: A Deep Learning Approach", In: *Paradigms of Communication, Computing and Data Analytics. Algorithms for Intelligent Systems,* 2023. Singapore: Springer Nature, 2023, pp. 377–388.
[http://dx.doi.org/10.1007/978-981-99-4626-6_32]

[9] E.S. Arı, and C. Gencer, "The use and comparison of a deterministic, a stochastic, and a hybrid multiple-criteria decision-making method for site selection of wind power plants: An application in Turkey", *Wind Eng.,* vol. 44, no. 1, pp. 60-74, 2020. [http://dx.doi.org/10.1177/0309524X19849831]

[10] I. Ul Haq, "Diagnosis of Neurological Disease Using Bioinspired Algorithms", In: Bio-Inspired Optimization for Medical Data Mining, S. Srivastava, A. Anand, A. Kumar, B. Saini, P. S. Rathore, Eds. Hoboken, NJ, USA: Wiley-IEEE Press, 2024, pp. 245-270. [http://dx.doi.org/10.1002/9781394213894.ch11] [http://dx.doi.org/10.1002/9781394214211.ch12]

[11] X. Wu, C. Zhang, L. Jiang, and H. Liao, "An integrated method with PROMETHEE and conflict analysis for qualitative and quantitative decision-making: case study of site selection for wind power plants", *Cognit. Comput.,* vol. 12, no. 1, pp. 100-114, 2020. [http://dx.doi.org/10.1007/s12559-019-09675-7]

[12] P. Prajapati, R. Guo, A. Cai, and R. Prashad, "Navigating the energy transition in India: challenges and opportunities towards sustainable energy goal," *Water-Energy Nexus*, vol. 9, pp. 1–18, 2026.

[13] G. Rediske, H.P. Burin, P.D. Rigo, C.B. Rosa, L. Michels, and J.C.M. Siluk, "Wind power plant site selection: A systematic review", *Renew. Sustain. Energy Rev.,* vol. 148, p. 111293, 2021. [http://dx.doi.org/10.1016/j.rser.2021.111293]

[14] D. Song, G. Shen, C. Huang, Q. Huang, J. Yang, M. Dong, Y.H. Joo, and N. Duić, "Review on the application of artificial intelligence methods in the control and design of offshore wind power systems", *J. Mar. Sci. Eng.,* vol. 12, no. 3, p. 424, 2024. [http://dx.doi.org/10.3390/jmse12030424]

[15] I. U. Haq, and A. M. U. D. Khanday, "Exploring the Role of Explainable AI in Women's Health: Challenges and Solutions", In: Explainable Artificial Intelligence in the Healthcare Industry, A. Kumar, T. A. Kumar, P. Das, C. Sharma, A. K. Dubey, Eds. Hoboken, NJ, USA: Wiley-IEEE Press, 2025, pp. 283–306. [http://dx.doi.org/10.1002/9781394249312.ch13]

[16] A. Amsharuk, and G. Łaska, "The approach to finding locations for wind farms using GIS and MCDA: Case study based on podlaskie voivodeship, Poland", *Energies,* vol. 16, no. 20, p. 7107, 2023. [http://dx.doi.org/10.3390/en16207107]

[17] G. Placide, and M.R. Lollchund, "Wind farm site selection using GIS-based mathematical modeling and fuzzy logic tools: a case study of Burundi", *Front. Energy Res.,* vol. 12, p. 1353388, 2024. [http://dx.doi.org/10.3389/fenrg.2024.1353388]

[18] T. Mollick, G. Hashmi, and S.R. Sabuj, "Wind speed prediction for site selection and reliable operation of wind power plants in coastal regions using machine learning algorithm variants", *Sustainable Energy Research,* vol. 11, no. 1, p. 5, 2024. [http://dx.doi.org/10.1186/s40807-024-00098-z]

[19] A.N. Petrov, and J.M. Wessling, "Utilization of machine-learning algorithms for wind turbine site suitability modeling in Iowa, USA", *Wind Energy (Chichester Engl.),* vol. 18, no. 4, pp. 713-727, 2015. [http://dx.doi.org/10.1002/we.1723]

[20] B. Liu, X. Ma, J. Guo, H. Li, S. Jin, Y. Ma, and W. Gong, "Estimating hub-height wind speed based on a machine learning algorithm: implications for wind energy assessment", *Atmos. Chem. Phys.,* vol. 23, no. 5, pp. 3181-3193, 2023. [http://dx.doi.org/10.5194/acp-23-3181-2023]

[21] Y. Yang, H. Lou, J. Wu, S. Zhang, and S. Gao, "A survey on wind power forecasting with machine learning approaches", *Neural Comput. Appl.,* vol. 36, no. 21, pp. 12753-12773, 2024. [http://dx.doi.org/10.1007/s00521-024-09923-4]

[22] Y.K. Wu, C.Y. Lee, C.R. Chen, K.W. Hsu, and H.T. Tseng, "Optimization of the wind turbine layout

and transmission system planning for a large-scale offshore windfarm by AI technology", *IEEE Trans. Ind. Appl.,* vol. 50, no. 3, pp. 2071-2080, 2014. [http://dx.doi.org/10.1109/TIA.2013.2283219]

[23] P.D. Cerna, R.S. Evangelista, C.M. Castillo, J.A. Muallam-Darkis, M.A.C. Velasco, J.P. Legaspi, and M.M. Gatdula, "Wind Power Plant Site Selection using Integrated Machine Learning and Multiple-Criteria Decision Making Technique", In: *E3S Web of Conferences.* vol. 405. EDP Sciences, 2023, p. 02030. [http://dx.doi.org/10.1051/e3sconf/202340502030]

[24] Ehsan, M. A. (2021). Predictive models for wind speed using artificial intelligence and copula. arXiv preprint arXiv:2107.06182.

[25] S. Ashwin Renganathan, R. Maulik, S. Letizia, and G.V. Iungo, "Data-driven wind turbine wake modeling via probabilistic machine learning", *Neural Comput. Appl.,* vol. 34, no. 8, pp. 6171-6186, 2022. [http://dx.doi.org/10.1007/s00521-021-06799-6]

[26] J.F. Manwell, J.G. McGowan, and A.L. Rogers, *Wind energy explained: theory, design and application.* John Wiley & Sons, 2010.

[27] T. Ackermann, Ed., *Wind power in power systems.* John Wiley & Sons, 2012. [http://dx.doi.org/10.1002/9781119941842]

[28] D. Toke, "Explaining wind power planning outcomes: Some findings from a study in England and Wales", *Energy Policy,* vol. 33, no. 12, pp. 1527-1539, 2005. [http://dx.doi.org/10.1016/S0301-4215(04)00017-5]

[29] M.I. Blanco, "The economics of wind energy", *Renew. Sustain. Energy Rev.,* vol. 13, no. 6-7, pp. 1372-1382, 2009. [http://dx.doi.org/10.1016/j.rser.2008.09.004]

[30] S.C. Pryor, and R.J. Barthelmie, "Climate change impacts on wind energy: A review", *Renew. Sustain. Energy Rev.,* vol. 14, no. 1, pp. 430-437, 2010. [http://dx.doi.org/10.1016/j.rser.2009.07.028]

[31] P. Devine-Wright, "Beyond NIMBYism: towards an integrated framework for understanding public perceptions of wind energy", *Wind Energy (Chichester Engl.),* vol. 8, no. 2, pp. 125-139, 2005. [http://dx.doi.org/10.1002/we.124]

[32] R. H. Wiser and M. Bolinger, *2018 Wind Technologies Market Report,* 2019.

[33] S.A. Kalogirou, "Artificial neural networks in renewable energy systems applications: a review", *Renew. Sustain. Energy Rev.,* vol. 5, no. 4, pp. 373-401, 2001. [http://dx.doi.org/10.1016/S1364-0321(01)00006-5]

[34] C. Cortes, and V. Vapnik, "Support vector networks", *Mach. Learn.,* vol. 20, no. 3, pp. 273-297, 1995. [http://dx.doi.org/10.1023/A:1022627411411]

[35] L. Breiman, "Random Forests", *Mach. Learn.,* vol. 45, no. 1, pp. 5-32, 2001. [http://dx.doi.org/10.1023/A:1010933404324]

[36] J.H. Holland, *Adaptation in natural and artificial systems: an introductory analysis with applications to biology, control, and artificial intelligence.* MIT press, 1992. [http://dx.doi.org/10.7551/mitpress/1090.001.0001]

[37] L.A. Zadeh, "Fuzzy sets", *Inf. Control,* vol. 8, no. 3, pp. 338-353, 1965. [http://dx.doi.org/10.1016/S0019-9958(65)90241-X]

[38] Y. LeCun, Y. Bengio, and G. Hinton, "Deep learning", *Nature,* vol. 521, no. 7553, pp. 436-444, 2015. [http://dx.doi.org/10.1038/nature14539]

[39] L. Bai, Y. Yang, L. Zhou, Y. Li, Y. Xiao, and W. Shi, "Optimal design and performance improvement of an electric submersible pump impeller based on Taguchi approach", *Energy,* vol. 252, p. 124032,

2022.
[http://dx.doi.org/10.1016/j.energy.2022.124032]

[40] I. Blanco, G. Vox, E. Schettini, and G. Russo, "Assessment of the environmental loads of green façades in buildings: a comparison with un-vegetated exterior walls", *J. Environ. Manage.*, vol. 294, p. 112927, 2021.
[http://dx.doi.org/10.1016/j.jenvman.2021.112927] [PMID: 34107375]

[41] J.W. Ni, M.J. Li, and T. Ma, "The study of energy filtering management process for microgrid based on the dynamic response model of vanadium redox flow battery", *Appl. Energy,* vol. 336, p. 120867, 2023.
[http://dx.doi.org/10.1016/j.apenergy.2023.120867]

[42] P.P. Wang, Y.P. Li, G.H. Huang, and S.G. Wang, "A multivariate statistical input–output model for analyzing water-carbon nexus system from multiple perspectives - Jing-Jin-Ji region", *Appl. Energy,* vol. 310, p. 118560, 2022.
[http://dx.doi.org/10.1016/j.apenergy.2022.118560]

[43] C. Gualtieri, and H. Chanson, "Physical and numerical modelling of air-water flows: An Introductory Overview", *Environ. Model. Softw.,* vol. 143, p. 105109, 2021.
[http://dx.doi.org/10.1016/j.envsoft.2021.105109]

[44] A. Raihan, "A comprehensive review of artificial intelligence and machine learning applications in energy sector", *J. Technol. Innov. Energy,* vol. 2, no. 4, pp. 1-26, 2023.
[http://dx.doi.org/10.56556/jtie.v2i4.608]

[45] L.A. Yousef, H. Yousef, and L. Rocha-Meneses, "Artificial intelligence for management of variable renewable energy systems: a review of current status and future directions", *Energies,* vol. 16, no. 24, p. 8057, 2023.
[http://dx.doi.org/10.3390/en16248057]

[46] W. Guan, X. Chen, J. Zhang, H. Hu, and C. Liang, "Catalytic transfer hydrogenolysis of lignin α-O-4 model compound 4-(benzyloxy)phenol and lignin over Pt/HNbWO6/CNTs catalyst", *Renew. Energy,* vol. 156, pp. 249-259, 2020.
[http://dx.doi.org/10.1016/j.renene.2020.04.078]

[47] M. Blonsky, J. Maguire, K. McKenna, D. Cutler, S.P. Balamurugan, and X. Jin, "OCHRE: The object-oriented, controllable, high-resolution residential energy model for dynamic integration studies", *Appl. Energy,* vol. 290, p. 116732, 2021.
[http://dx.doi.org/10.1016/j.apenergy.2021.116732]

[48] M.T. Ribeiro, S. Singh, and C. Guestrin, "Why should i trust you?" Explaining the predictions of any classifier", In: *Proceedings of the 22nd ACM SIGKDD International Conference on Knowledge Discovery and Data Mining.* San Francisco, CA, USA, pp. 1135–1144.
[http://dx.doi.org/10.18653/v1/N16-3020]

[49] Y. Wu, J. Rong, D. Wang, X. Zhao, L. Meng, M. Arıcı, C. Liu, R. Yang, and D. Li, "Synergistic enhancement of heat transfer and thermal storage characteristics of shell and tube heat exchanger with hybrid nanoparticles for solar energy utilization", *J. Clean. Prod.,* vol. 387, p. 135882, 2023.
[http://dx.doi.org/10.1016/j.jclepro.2023.135882]

SUBJECT INDEX

A

B

C

D

E

Sugandha Singh, Inam Ul Haq, Sachin Ahuja, Akib Mohi Ud Din Khanday & Neha Sharma (Eds.)

F

G

H

I

K

M

O

P

R

S

T

U

V

W

X

Y

www.ingramcontent.com/pod-product-compliance
Lightning Source LLC
LaVergne TN
LVHW070948210926
848482LV00002B/135